TI-83, TI-83 Plu
TI-83 Silver Edition
Graphing Calculator
Manual

Kathleen McLaughlin • Dorothy Wakefield

STATISTICS
Informed Decisions
Using Data

Michael Sullivan III

PEARSON

Prentice
Hall

Upper Saddle River, NJ 07458

Editor-in-Chief: Sally Yagan
Supplement Editor: Joanne Wendelken
Assistant Managing Editor: John Matthews
Production Editor: Wendy A. Perez
Supplement Cover Manager: Paul Gourhan
Supplement Cover Designer: Joanne Alexandris
Manufacturing Buyer: Ilene Kahn

Printed in the United States of America

10 9 8 7 6 5 4 3 2 1

ISBN 0-13-046496-1

Pearson Education Ltd., *London*
Pearson Education Australia Pty. Ltd., *Sydney*
Pearson Education Singapore, Pte. Ltd.
Pearson Education North Asia Ltd., *Hong Kong*
Pearson Education Canada, Inc., *Toronto*
Pearson Educación de Mexico, S.A. de C.V.
Pearson Education—Japan, *Tokyo*
Pearson Education Malaysia, Pte. Ltd.
Pearson Education, *Upper Saddle River, New Jersey*

▶ Introduction

The TI-83, TI-83 Plus and TI-83 Silver Edition Graphing Calculator Manual is one of a series of companion technology manuals that provide hands-on technology assistance to users of Sullivan *Statistics: Informed Decisions Using Data.*

Detailed instructions for working selected examples and problems from *Statistics: Informed Decisions Using Data* are provided in this manual. To make the correlation with the text as seamless as possible, the table of contents includes page references for both the Sullivan text and this manual.

▶ Contents:

Getting Started with the TI-83 Graphing Calculators

▶ Overview

This manual is designed to be used with the TI-83, the TI-83 Plus and the TI-83 Silver Edition Graphing Calculators. These calculators have a variety of useful functions for doing statistical calculations and for creating statistical plots. The commands for using the statistical functions are the same for all three calculators. The main difference among the three calculators is that the TI-83 Plus and the TI-83 Silver Edition can receive a variety of software applications that are available through the TI website (www.ti.com). TI also will provide downloadable updates to the TI-83 Plus and TI-83 Silver Edition operating systems. These features are not available on the TI-83.

Your textbook comes with data files on the CD data disk that can be loaded onto all three of the TI-83 calculators. In order to transfer data from a computer (IBM compatible) to the TI-83 or TI-83 Plus, you must purchase a Graph Link from Texas Instruments which connects the calculator to the computer. The TI-83 Silver Edition comes with the TI-Graph Link.

Throughout this manual all references to the TI-83 are actually references to the TI-83, TI-83 Plus and TI-83 Silver Edition calculators.

Before you begin using the TI-83, spend a few minutes becoming familiar with its basic operations. First, notice the different colored keys on the calculator. The gray keys are the number keys. The blue keys along the right side of the keyboard are the common math functions. The blue keys across the top set up and display graphs. The primary function of each key is printed in white on the key. For example, when you press STAT, the STAT MENU is displayed.

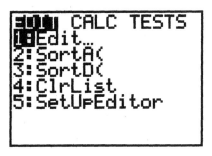

The secondary function of each key is printed in yellow above the key. When you press the 2nd key, the function printed in yellow above the key becomes active and the cursor changes from a solid rectangle to an ↑ (up-arrow). For example, when you press 2nd and the $\boxed{x^2}$ key, the $\sqrt{}$ function is activated. The notation used in this manual to indicate a secondary function is 2nd followed by the name of the secondary function. For example, to use the LIST function, found above the STAT key, the notation used in this manual is 2nd [LIST]. The LIST MENU will then be activated and displayed on the screen.

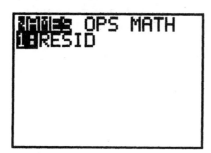

The alpha function of each key is printed in green above the key. When you press the green ALPHA key, the function printed in green above the key is activated and the cursor changes from a solid rectangle to A.

In this manual you will find detailed explanations of the different statistical functions that are programmed into the TI-83. These explanations will accompany selected examples from your textbook. This will give you the opportunity to learn the various calculator functions as they apply to the specific statistical material in each chapter.

▶ Getting Started

To operate the calculator, press ON in the lower left corner of the calculator. Begin each example with a blank screen, with a rectangular cursor flashing in the upper left corner. If you turn on your calculator and you do not have a blank screen, press the CLEAR key. You may have to press CLEAR a second time in order to clear the screen. If using the CLEAR key does not clear the screen, you can push 2nd [QUIT] (Note: QUIT is found above the MODE key.)

▶ Helpful Hints

To adjust the display contrast, push and release the 2nd key. Then push and hold the blue up arrow ▲ to darken or the blue down arrow ▼ to lighten.

The calculator has an automatic turn off that will turn the calculator off if it has been idle for several minutes. To restart, simply press the ON key.

There are several different graphing techniques available on the TI-83. If you inadvertently leave a graph on and attempt to use a different graphing function, your graph display may be cluttered with extraneous graphs, or you may get an ERROR message on the screen.

There are several items that you should check before graphing anything. First, press the Y= key, found in the upper left corner of the key pad, and clear all the Y-variables. The screen should look like the following display:

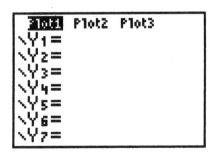

If there are any functions stored in the Y-variables, simply move the cursor to the line that contains a function and press CLEAR ENTER.

Next, press 2nd [STAT PLOT] (found on the Y= key) and check to make sure that all the STAT PLOTS are turned OFF.

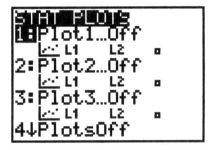

If you notice that a Plot is turned **ON**, select the Plot by using the blue down arrow key to highlight the number to the left of the Plot , press ENTER and move the cursor to **OFF** and press ENTER. Press 2nd [QUIT] to return to the home screen.

Data Collection

Section 1.2

▶ Example 4 (pg. 18) Generating a Simple Random
                      Sample

The first step is to set the *seed* by selecting a 'starting number' and storing this
number in **rand**. Suppose, for this example, that we select the number '34' as the
starting number. Type **34** into your calculator and press the STO key found on
the lower left section of the calculator keys. Next press the MATH key found in
the upper left section of the calculator keys. The Math Menu will appear.

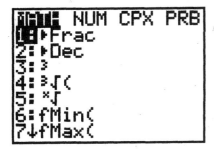

Use the blue right arrow key, ▶ , to move the cursor to highlight **PRB**. The
Probability Menu will appear.

```
MATH NUM CPX PRB
1:rand
2:nPr
3:nCr
4:!
5:randInt(
6:randNorm(
7:randBin(
```

The first selection on the **PRB** menu is **rand,** which stands for 'random number'. Notice that this highlighted. Simply press ENTER and the starting value of '42' will be stored into **rand** and will be used as the *seed* for generating random numbers.

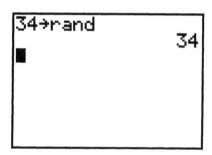

```
34→rand
            34
■
```

Now you are ready to generate a random integer. Press MATH again and the Math Menu will appear. Use the blue right arrow key, ▶ , to move the cursor to highlight **PRB**. The Probability Menu will appear. Select **5:RandInt(** by using the blue down arrow key, ▼ , to highlight it and pressing ENTER or by pressing the 5 key. **RandInt(** should appear on the screen. This function requires two values: the starting integer, followed by a comma (the comma is found on the black key above the 7 key), and the ending integer.   Close the parentheses and press ENTER. (Note: It is optional to close the parenthesis at the end of the command.) This command will generate one random number.

Now here is an example. In this example, one random number between 1 and 15 has been generated. The random number is 2.

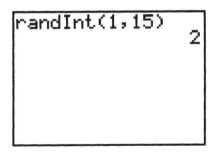

If you want to generate several random numbers with one command, you can change the **Randint** command so that it contains three values: the starting value, the ending value and the number of values you want to generate.

For an example, suppose you want to generate 15 values from the integers ranging from 1 to 50. The command is **randInt(1,50,15)**.

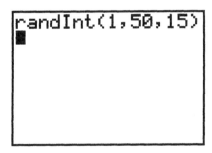

Press **ENTER** and a partial display of the 15 random integers should appear on your screen. (Note: your numbers will probably be different from the ones you see here. The numbers that are generated will depend on the *seed* that is initially selected.)

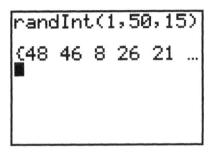

Use the right arrow to scroll through your 15 items. You might find that you have some duplicate values. The TI-83 uses a method called "sampling with replacement" to generate random numbers. This means that it is possible to select the same integer twice.

In the example in the text, you are asked to select a random sample of 10 residents from the 8,791 residents in the Village of Lemont. One way to choose the sample is to number the residents from 1 to 8,791 and randomly select 10 different residents. This sampling process is "without replacement." Since the TI-83 samples "with replacement", duplicates may occur in the sample. The best way to obtain 10 different residents is to generate more than 10 random integers, and to select the first 10 different numbers from the list. To be safe, you should generate 15 random integers.

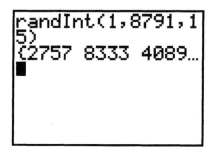

Use the right arrow to scroll to the right to see the rest of the list and write down the first 10 distinct values. Note: The numbers you generate depend on the *seed* and will differ from the numbers in this example.

▶ Problem 15 (pg.19) Generating a Simple Random Sample

a.   To obtain a simple random sample of size 10, press **MATH**, use the blue
     right arrow to highlight **PRB** and select **5: randInt**. Enter the starting
     value of **1**, the ending value of **50** and a sample size of **15**. We are
     generating more numbers than we actually need because of the
     possibility of getting duplicates in the sample. To obtain your sample of
     10, select the first 10 distinct numbers in the sample that you generated.
     (Note: In this example, we did not set a new seed. Setting a seed is
     optional. It is not required.)

b.   Repeat the steps in 15a and generate another sample of integers.

Organizing and Summarizing Data

CHAPTER

2

Section 2.2

▶ Example 2 (pg. 68) A histogram for Discrete data

To create a histogram, you have two choices: 1): enter all the individual data points from Table 8 on pg. 68 into one column or 2): enter the data values into one column and the frequencies into another column using Table 9 on pg. 68. For this example, we will use the table.

To create this histogram, you must enter information into List1 (**L1**) and List 2 (**L2**) on your calculator. You will enter the 'number of customers' into **L1** and the frequencies into **L2**. Press **STAT** and the Statistics Menu will appear.

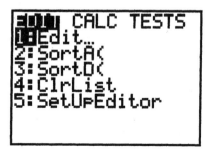

Press **ENTER** and lists **L1**, **L2** and **L3** will appear.

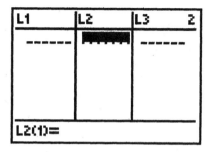

If the lists already contain data, you should clear them before beginning this example. Move your cursor so that the List name (**L1, L2,** or **L3**) of the list that contains data is highlighted.

```
| ■1      |L2      |L3      1|
| 1       |4       |------   |
| 2       |6       |         |
| 3       |8       |         |
| 10      |        |         |
| ------  |        |         |
|         |        |         |
| L1 ={1,2,3,10}              |
```

Press **CLEAR** **ENTER**. Repeat this process until all three lists are empty.

```
| L1      |L2      |L3      1|
| ▬▬▬▬▬  |4       |------   |
|         |6       |         |
|         |8       |         |
|         |------  |         |
|         |        |         |
|         |        |         |
| L1(1) =                     |
```

To enter the data values into **L1,** move your cursor so that it is positioned in the 1st position in **L1.** Type in the first value, **1,** and press **ENTER** or use the blue down arrow. Enter the next value, **2.** Continue this process until all 11 data values are entered into **L1**. Now use the blue up-arrow to scroll to the top of **L1.** As you scroll through the data, check it. If a data point is incorrect, simply move the cursor to highlight it and type in the correct value. When you have moved to the 1st value in **L1,** use the right arrow to move to the first position in **L2**. Enter the frequencies into **L2.**

```
| L1      |L2      |L3      3|
| 1       |1       |▬▬▬▬▬  |
| 2       |6       |         |
| 3       |1       |         |
| 4       |4       |         |
| 5       |7       |         |
| 6       |11      |         |
| 7       |5       |         |
| L3(1)=                      |
```

Before graphing the histogram, make sure that there are no functions in the Y-registers. To do this, press the **Y=** key. If there are any functions stored in any of the Y-values, simply move the cursor to the line that contains a function and press CLEAR . Now you are ready to graph the histogram. Press **2ⁿᵈ** [STAT PLOT] (located above the Y= key).

Select Plot1 by pressing ENTER.

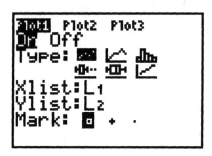

Notice that Plot1 is highlighted. On the next line, notice that the cursor is flashing on **ON** or **OFF**. Position the cursor on **ON** and press ENTER to select it. The next two lines on the screen show the different types of graphs. Move your cursor to the symbol for histogram (3ʳᵈ item in the 1ˢᵗ line of **Type**) and press ENTER.

The next line is **Xlist**. Use the blue down arrow to move to this line. On this line, you tell the calculator where the data values are stored. In most graphing situations, the data are entered into **L1** so **L1** is the default option. Notice that the cursor is flashing on **L1**. Push ENTER to select **L1**. The last line is the frequency line. On this line, **1** is the default. The cursor should be flashing on **1**. Change **1** to **L2** by pressing **2ⁿᵈ** [L2]. (Note: L2 is found above the 2 key.)

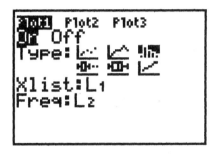

To view a histogram of the data, press ZOOM.

There are several options in the Zoom Menu. Using the blue down arrow, scroll down to option 9, **ZoomStat,** and press ENTER. A histogram should appear on the screen.

This histogram is not exactly the same as the ones on pg. 69 of your textbook. You can adjust the histogram so that it does look exactly like the one in your text. Press Window and set **Xmin** to 1, **Xmax** to 12 (this one extra data value is needed to complete the last bar of the histogram), and **Xscl** equal to 1, which is the difference between successive data values in the frequency distribution. Note: In many cases it is not necessary to change the values for **Ymin, Ymax** or **Yscl**. What you must do is to check these values and make sure that **Ymin** is a small negative value (a value between −5 and −1 would be good) and **Ymax** must be larger than the largest frequency value in your dataset. You never need to adjust **Yscl**.

```
WINDOW
 Xmin=1
 Xmax=12
 Xscl=1▉
 Ymin=-5.41242
 Ymax=21.06
 Yscl=1
 Xres=1
```

Press GRAPH.

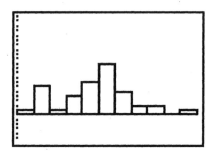

Notice the blue TRACE key. If you press it, a flashing cursor, ✳, will appear at the top of the 1st bar of the histogram.

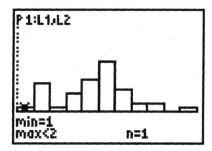

Notice the information at the bottom of the screen. **Min** is the actual data value for the first bar of the histogram. In this example, the first data value is **1**. We do not actually need to use the Max value in this example. "n=1" tells us that there is only one data point in the dataset that has a value of **1**. You can use the blue right arrow to move through each of the bars. For example, if you move to the 5th bar in the histogram, you will see that the data value for that bar is **5** and that there are 7 data points in the dataset that have a value of **5**.

Now that you have completed this example, turn Plot1 **OFF**. Using **2nd** [STAT PLOT], select Plot1 by pressing ENTER and highlighting **OFF**. Press ENTER and **2nd** [QUIT]. (Note: Turning Plot1 **OFF** is optional. You can leave it ON but leaving it ON will effect other graphing operations of the calculator.)

◀

▶ Example 4 (pg. 72) A histogram for continuous data

Press **STAT** and **ENTER** to select **1:Edit**. If there is data in L1, highlight **L1** at the top of the first list and press **CLEAR** and **ENTER** to clear the data. You should also clear **L2**.

To create this histogram, use Table 13 on page 71. You must enter the midpoints of each class into List1 (**L1**) and the frequencies into List 2 (**L2**). To obtain the midpoints of each class, add the lower limit plus the upper limit and divide by 2. For example, here is the calculation for the first class: (10.0+14.9)/2= 12.45.

To enter the midpoints in L1, you can do the calculation for the midpoints right on this screen. Simply type the calculation on the data entry line and push **ENTER** . The calculation will be automatically converted to the midpoint.

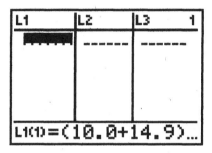

To set up the histogram, push **2ⁿᵈ [STAT PLOT]** and **ENTER** to select **Plot 1**. Turn ON **Plot 1**, set **Type** to **Histogram**, set **Xlist** to **L1,**. set **Freq** to **L2**.

Press **ZOOM** and **9:ZoomStat** and press **ENTER** and a histogram will appear on the screen. Press **Window** to adjust the Graph Window. Set **Xmin** equal to 10.0 (the lower limit of the first class) and **Xmax** equal to 50.0 (a value that would be the lower limit of an additional class at the end of the table. This extra value is needed to complete the last bar of the histogram). Set **Xscl** equal to 5, which is the class width. (Note: In many cases it is not necessary to change the values for **Ymin**, **Ymax** or **Yscl**. What you must do is to check these values and make sure that **Ymin** is a small negative value (a value between –5 and –1 would be good) and **Ymax** must be larger than the largest frequency value in your dataset. You never need to adjust **Yscl**.

Press GRAPH and the histogram should appear.

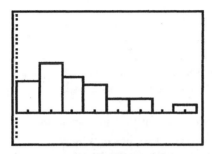

You can press TRACE and scroll through the bars of the histogram. The minimum value of the class will appear as **Min.** **Max** is written as an inequality that states that the maximum value in the class is *less than* the given value. **n** is the number of data points in the class.

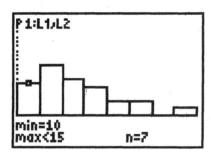

Notice, for example, with the cursor highlighting the first bar of the histogram, you will see that the first class contains values greater than or equal to 10 and less than 15 and that there are 7 data points in the this class.

> ▶ Problem 21 (pg. 81)

For this example, we will construct the histogram first (part c.) and then use it to fine the frequencies for the frequency distribution (part a.).

Press STAT and select **1:Edit** and press ENTER. Highlight the name "L1" and press CLEAR and ENTER. You can also clear L2 but you will not be using **L2** in this example. Enter the data values into **L1**.

To set up the histogram, push **2ⁿᵈ [STAT PLOT]** and ENTER to select **Plot 1**. Turn ON **Plot 1**, set **Type** to **Histogram**, set **Xlist** to **L1**. In this example, you must set **Freq** to 1. If the frequency is set on **L2** move the cursor so that it is flashing on **L2** and press CLEAR. The cursor is now in ALPHA mode (notice that there is an "A" flashing in the cursor). Push the ALPHA key and the cursor should return to a solid flashing square. Type in the number **1**.

Press Window to set the Graph Window. The first value you must enter is the value for **Xmin**. This value will be the lower class limit of the first class which is **10**. The value for **Xmax** would be the lower class limit of the one extra class that would be needed to complete the last bar of the histogram. Look through the data in your textbook. Notice that the largest data point is 73, therefore, the last class would be 70-79. The lower class limit of the next class would be **80**. This is the value for **Xmax**. Set **Xscl** equal to 10, which is the class width. (Note: In many cases it is not necessary to change the values for **Ymin**, **Ymax** or **Yscl**. What you must do is to check these values and make sure that **Ymin** is a small negative value (a value between −5 and −1 would be good) and **Ymax** must be larger than the largest frequency value in your dataset. You never need to adjust **Yscl**.

Press GRAPH and the histogram should appear.

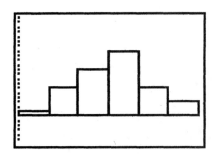

You can press ▣TRACE▣ and scroll through the bars of the histogram.
The minimum value of the class will appear as **Min.** **Max** is written as an
inequality that states that the maximum value in the class is *less than* the given
value. **n** is the number of data points in the class.

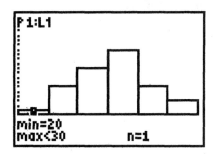

In the screen shown here, you see that the first class is 20-29 and there is one data
point in this class. Use the blue right arrow to scroll through the bars of the
histogram and use this information to construct the frequency distribution for part
(a) of this problem.

Here is a starting setup for the frequency distribution table:

Class	Frequency	Relative Frequency
20-29	1	
30-39		

Enter the remaining classes. Now go through the dataset and complete the
frequency column by recording the number of values in each class. To complete
the relative frequency column, simply divide each frequency by the total
frequency. For example, the relative frequency for the first class would be 1/40.

To do part (f) of the problem, press Window to set the Graph Window. Set Xscl equal to 5. Press GRAPH and the histogram should appear.

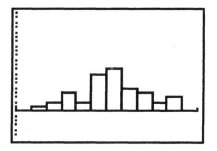

Section 2.3

▶ Example (pg. 85-86) A Frequency Polygon

Press **STAT** and select **1:Edit** and press **ENTER**. Clear all data from **L1** and **L2**. Enter the midpoints from Table 17 on pg. 86 into **L1** and enter the frequencies into **L2**.

To set up the frequency polygon, press 2nd **[STAT PLOT]** . Press **ENTER** to select **Plot 1**. Highlight **On** and press **ENTER**. Set **Type** to the frequency polygon which is the second selection and press **ENTER**. Set **Xlist** to **L1** and **Freq** to **L2**. Next, there are three different types of **Marks** that you can select for the graph. The first choice, a small square, is the best one to use.

Press **ZOOM** and select **9:ZoomStat** and **ENTER**.

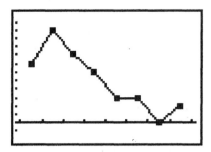

You can press **TRACE** and scroll through the points in the polygon. For example, the third data point represents class 3 which has a midpoint of 22.45 and a frequency of 8.

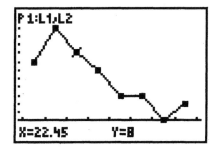

▶ Example (pg. 87) An ogive

Press **STAT** and select **1:Edit** and press **ENTER**. Clear all data from **L1** and **L2**. Using the data in Table 18 on pg. 87, enter the upper class limit into **L1** and enter the cumulative frequencies into **L2**.

L1	L2	L3	2
14.9	7	------	
19.9	18		
24.9	26		
29.9	32		
34.9	35		
39.9	36		
44.9	38		

L2(6) =38

To set up the ogive, press **2ⁿᵈ [STAT PLOT]** . Press **ENTER** to select **Plot 1**. Highlight **On** and press **ENTER**. Set **Type** to the frequency polygon which is the second selection and press **ENTER**. Set **Xlist** to **L1** and **Freq** to **L2**.

Press **ZOOM** and select **9:ZoomStat** and **ENTER**.

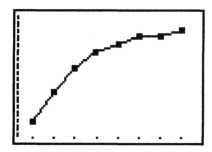

You can press **TRACE** and scroll through the points in the ogive.

▶ Example 1 (pg. 88)   A Time Series Plot

Press **STAT** and select **1:Edit**. Clear **L1** and **L2**. Notice that the dates in Table 19 on pg. 88 are "January 2000" through "November 2001." Rather than entering these actual dates into **L1**, you can simply number the months from "1" to "23" and enter these numbers into **L1**. Enter the closing prices into **L2**.

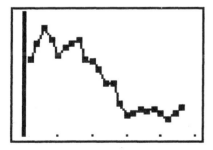

L1	L2	L3 2
1	54.75	------
2	66.094	
3	77.313	
4	69.328	
5	56.938	
6	63.563	
7	65.438	

L2(1)=54.75

Notice that the values displayed in **L2** have been rounded to the thousandths place. This is done automatically on the calculator. The actual values are still stored in the calculator; the rounded values are simply for the display in **L2**.

To construct the time series chart, press **2ⁿᵈ [STAT PLOT]** and select **1:Plot 1** and **ENTER**. Turn ON **Plot 1**. Set the **Type** to **frequency polygon**. For **Xlist** select **L1** and for **Ylist** select **L2**.

Press **ZOOM** and scroll down to **9:ZoomStat** and press **ENTER** or simply press **9** and **ZoomStat** will automatically be selected.

Use **TRACE** to scroll through the data values for each year. Notice for example, the closing price for the 4ᵗʰ month is 69.3281.

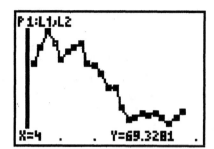

▶ Problem 9 (pg. 90)

For this exercise, use the dataset in Problem 21 on page 81.

a.) To construct a cumulative frequency distribution table, look through the dataset and find the minimum value (28) and the maximum value (73). Select "20" as a convenient starting number for your first class. Set the class width at 10. Here is a starting set up for the table.

Class	Midpoint	Frequency	Relative Frequency	Cumulative Frequency	Cumulative Relative Frequency
20-29					
30-39					

Enter the remaining classes and enter the midpoints of each class. Now go through the dataset and complete the frequency column by recording the number of values in each class. To complete the cumulative frequency column, simply accumulate the frequencies. For example, the cumulative frequency for the third class would be the sum of the frequencies for the 1st, 2nd and 3rd classes.

b.) Once you have completed the frequency column, simply divide each frequency by the total frequency (40) and enter these values in the relative frequency column. To complete the cumulative relative frequency column, simply accumulate the relative frequencies.

c.) Press STAT and select 1:Edit. Clear the lists and enter the midpoints into L1 and the frequencies into L2. Press 2nd [STAT PLOT] and select **Plot 1** and press ENTER. Set the **Type** to **frequency polygon**. Set **Xlist** to L1 and Freq to L2. Press GRAPH and the frequency polygon should appear.

d.) Press STAT and select 1:**Edit** and press ENTER. Clear all data from **L1 and L2.** Enter the upper class limits into L1 and enter the cumulative frequencies into L2. Press 2nd [STAT PLOT] and select **Plot 1** and press ENTER. Set the **Type** to **frequency polygon**. Set **Xlist** to L1 and Freq to L2. Press GRAPH and the frequency ogive should appear.

e.) A relative frequency ogive is constructed using the upper class limits and the relative frequencies (rather than the frequencies). The actual picture on the TI-83 would be identical to the frequency ogive that you constructed in part (d).

Numerically Summarizing Data

CHAPTER

3

Section 3.1

▶ Example 1 (pp. 113)   A population mean and a sample mean

The TI-83 has one method for calculating the mean of a dataset. This method is used for a population mean, μ, and a sample mean, $\bar{x}$. The calculator always used $\bar{x}$, the symbol for the sample mean.

(a). Press **STAT** and select **1:Edit**. Clear **L1** and enter the data into **L1**. Press **STAT** again and highlight **CALC** to view the Calc Menu.

```
EDIT CALC TESTS
1∎1-Var Stats
2:2-Var Stats
3:Med-Med
4:LinReg(ax+b)
5:QuadReg
6:CubicReg
7↓QuartReg
```

Select **1:1-Var Stats**. On this line, enter the name of the column that contains the data. Since you have stored the data in **L1**, simply enter **2ⁿᵈ** **[L1]** **ENTER** and the first page of the one variable statistics will appear. (Note: If you did not enter a column name, the default column, which is **L1,** would be automatically selected.)

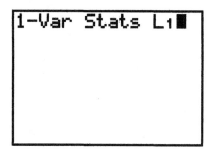

```
1-Var Stats L1∎



```

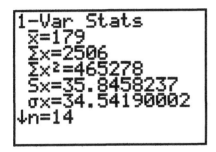

The first item is the mean of the dataset. The correct symbol for the mean of a population is μ. For this dataset, $\mu = 179$.

(b). To use a seed of '55', press **55** STO MATH and select **PRB**. Press ENTER to select **1:Rand** and press ENTER. (Note: Selecting a seed is an optional step which can be omitted when generating random data.)

To generate a random sample of 5 teams from the 14 teams, press MATH and select **PRB**. Select **5:RandInt** by pressing **5** or moving the cursor to **5:RandInt**. Enter a starting value of **1**, an ending value of **14** and a sample size of **7**. (Recall: The TI-83 samples with replacement. This method may result in duplicates in your sample. Selecting a few more values than you need will allow you to skip duplicate values.)

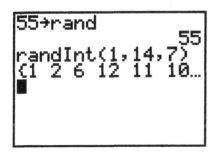

In this outcome there were no duplicates. The random sample of 5 teams would be the first 5 numbers: 1,2,6,12 and 11. (Note: If you had not selected '55' as the seed, you would have obtained a different set of random numbers.)

(c.) Press STAT and select **1:Edit**. Clear **L2** and enter the 5 data values into **L2**. Press STAT again and highlight **CALC** to view the Calc Menu. Select **1:1-Var Stats**. On this line, enter the name of the column that contains the data. Since you have stored the data in **L2**, simply enter **2ⁿᵈ** **[L2]** ENTER and the first page of the one variable statistics will appear.

```
1-Var Stats
 x̄=144.6
 Σx=723
 Σx²=105983
 Sx=18.95521037
 σx=16.95405556
↓n=5
```

The sample mean, $\bar{x}$, is 144.6.

▶ Example 2 (pg. 116) The median of a dataset

Press **STAT** and select **1:Edit**. Clear **L1** and enter the homerun data from Table 1 on pg. 114 into **L1**. Press **STAT** again and highlight **CALC** to view the Calc Menu. Select **1:1-Var Stats** and press **2nd** **[L1]** **ENTER**. Notice the down arrow in the bottom left corner of the screen. This indicates that more information follows this first page. Use the blue down arrow to scroll through this information. The third item you see on the second page is the median, Med = 182.

```
1-Var Stats
↑Sx=35.8458237
 σx=34.54190002
 n=14
 minX=121
 Q₁=152
↓Med=182
■
```

(Note: It is not necessary to put the data in ascending order when calculating a median using the TI-83.)

◀

▶ Problem 13 (pg. 125)

Press **STAT** and select **1:Edit**. Clear **L1** and enter the data into **L1**. Press
STAT again and highlight **CALC** to view the Calc Menu. Select **1:1-Var Stats**
and press 2nd **[L1]** **ENTER**.

(a.) The first item is the sample mean, 7.875. To find the median, scroll down to
the next page of output. The median is **8**.

(b.) Before graphing the histogram, make sure that there are no functions in the
Y-registers. To do this, press the **Y=** key. If there are any functions stored in any
of the Y-values, simply move the cursor to the line that contains a function and
press **CLEAR** . Now you are ready to graph the histogram. Press 2nd **[STAT
PLOT]** (located above the **Y=** key).

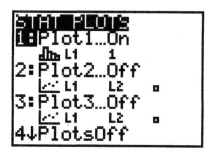

Select Plot1 by pressing **ENTER**. Position the cursor on **ON** and press **ENTER**
to select it. On the next line, move your cursor to the symbol for histogram (3rd
item in the 1st line of **Type**) and press **ENTER**. The next line is **Xlist**. Use the
blue down arrow to move to this line. On this line, you tell the calculator where
the data values are stored. In most graphing situations, the data are entered into
L1 so **L1** is the default option. Notice that the cursor is flashing on **L1**. Push
ENTER to select **L1**. The last line is the frequency line. On this line, **1** is the
default. The cursor should be flashing on **1**.

To view a histogram of the data, press ZOOM.

There are several options in the Zoom Menu. Using the blue down arrow, scroll down to option 9, **ZoomStat,** and press ENTER. A histogram should appear on the screen.

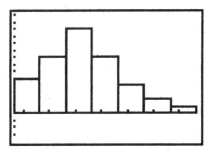

Section 3.2

▶ Example 5 (pg. 137) The Standard Deviation

Press **STAT** and select **1:Edit**. Clear **L1** and enter the population data from Table 8 on page 132 into **L1**. Press **STAT** and highlight **CALC** to display the Calc Menu. Select **1: 1-Var Stats** and press **2ⁿᵈ [L1] ENTER**. The population standard deviation is σx, 34.54190002.

```
1-Var Stats
 x̄=179
 Σx=2506
 Σx²=465278
 Sx=35.8458237
 σx=34.54190002
↓n=14
```

Press **STAT** and select **1:Edit**. Clear **L2** and enter the sample data points (158,136,139,121 and 169) into **L2**. Press **STAT** and highlight **CALC** to display the Calc Menu. Select **1: 1-Var Stats** and press **2ⁿᵈ [L2] ENTER**. The sample standard deviation is Sx, 18.95521037.

```
1-Var Stats
 x̄=144.6
 Σx=723
 Σx²=105983
 Sx=18.95521037
 σx=16.95405556
↓n=5
■
```

▶ Problem 11 (pg. 143)

(a.) Press **STAT** and select **1:Edit**. Clear **L1** and enter the data into **L1**. Press **STAT** and highlight **CALC** to display the Calc Menu. Select **1: 1-Var Stats** and press 2nd **[L1]** **ENTER**. The population standard deviation is σx, 7.67.

To calculate the population variance, $(\sigma x)^2$, type in the value of the standard deviation at the bottom of the screen and press the $\boxed{x^2}$ key and **ENTER**.

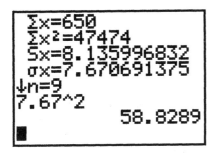

The population variance is 58.83.

(b.) To generate a random sample of size 3, first number the students from 1 to 9. The TI-83 samples with replacement so you will need to generate a sample that is larger than 3. Suppose you take a sample of size 5. Press **MATH** and **PRB**. Select **5:RandInt** and enter **1,9,5**. Here is one possible outcome.

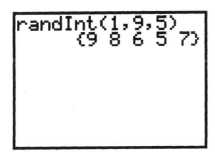

In this outcome we have selected 5 students: Student Numbers: 9,8,6,5 and 7. Since we need only need 3 students, simply select the first three: 9,8, and 6. (Note: Random samples vary so your results will be different.)

Enter the pulse rates of the 3 students into L2. Press **STAT** and highlight **CALC** to display the Calc Menu. Select **1: 1-Var Stats** and press 2nd **[L2]** **ENTER**. The sample standard deviation is Sx, 8.14.

To calculate the sample variance, $(Sx)^2$, type in the value of the standard deviation at the bottom of the screen and press the $\boxed{x^2}$ key and $\blacksquare ENTER$. The sample variance is 66.26.

Repeat this process to generate another random sample of size 3.

◀

▶ Problem 31 (pg. 147)

(a.) Press **STAT** and select **1:Edit**. Clear L1 and enter the sample data into **L1**. Press **STAT** and highlight **CALC** to display the Calc Menu. Select **1: 1-Var Stats** and press 2nd [L1] **ENTER**. The sample standard deviation, Sx, is 11.63.

(b.) Press **STAT** and select **1:Edit**. Clear **L2**. Move the cursor to the top of **L2** again and press **ENTER**. The cursor should now be flashing on the bottom line of the screen. On this line, type in 2nd [L1 + 4.

```
L1        L2        L3        2
65        ------    ------
70
71
75
95
------

L2 =L1+4
```

Press **ENTER**. Each value in **L2** should be 4 points higher than the corresponding value in **L1**. Press **STAT** and highlight **CALC** to display the Calc Menu. Select **1: 1-Var Stats** and press 2nd [L2] **ENTER**. The sample standard deviation, Sx, is 11.63, the same as the standard deviation for the original data.

(d.) Press **STAT** and select **1:Edit**. Clear **L3**. Move the cursor to the top of **L3** again and press **ENTER**. The cursor should now be flashing on the bottom line of the screen. On this line, type in 2nd [L1 *2. Press **STAT** and highlight **CALC** to display the Calc Menu. Select **1: 1-Var Stats** and press 2nd [L3] **ENTER**. The sample standard deviation, Sx, is 23.26, which is twice the value of the standard deviation for the original data.

◀

Section 3.3

▶ Example 1 (pg. 151)   The mean of a frequency distribution

Press **STAT** and select **1:Edit**. Clear **L1** and **L2**. Enter the midpoints of each class into **L1**. You can do the calculations for the midpoints directly on this screen. For the first class, type in **(10+14.9)/2** and press **ENTER**.

```
L1        L2        L3        1
━━━━━     ------    ------

L1(1)=(10+14.9)/2▋
```

The value of the midpoint, 12.45, will appear as the first entry in **L1**. Continue this process to obtain the midpoints for each of the classes. Enter the frequencies into **L2**. Press **STAT** and highlight **CALC** to display the Calc Menu. Select **1: 1-Var Stats** and press **2ⁿᵈ** **[L1]** **,** **2ⁿᵈ** **[L2]**. Press **ENTER**. (Note: You must place the comma between **L1** and **L2**).

```
1-Var Stats L₁,L
₂
```

Using **L1** and **L2** in the **1:1-Var Stats** calculation is necessary when approximating a mean from a frequency distribution.. The calculator uses the data in **L1** and the associated frequencies in **L2** to approximate the average for the dataset. In this example, the approximate mean is 23.2 percent.

◀

▶ Example 2 (pg. 152)   The weighted mean

Press **STAT** and select **1:Edit**. Clear **L1** and **L2**. Enter the point values for each letter grade that Marissa earned into **L1**. Enter the corresponding credits earned into **L2**.

```
L1      L2      L3      2
  4       4      ------
  3       3
  4       3
  2       5
  4       1
------  ------

L2(6) =
```

Press **STAT** and highlight **CALC** to display the Calc Menu. Select **1: 1-Var Stats** and press **2ⁿᵈ [L1] , 2ⁿᵈ [L2]**. Press **ENTER**. (Note: You must place the comma between **L1** and **L2**).

```
1-Var Stats L₁,L
2
```

```
1-Var Stats
 x̄=3.1875
 Σx=51
 Σx²=175
 Sx=.910585892
 σx=.8816709987
↓n=16
■
```

Her GPA (weighted average) is 3.1875.

▶ Example 3 (pg. 153) The Variance and Standard Deviation of
a Frequency Distribution

Press **STAT** and select **1:Edit**. Clear **L1** and **L2**. Enter the midpoints from
Table 12 on pg. 151 into **L1** and the frequencies into **L2**. Press **STAT**, highlight
CALC, select **1:1-Var Stats**, and press **2ⁿᵈ** **[L1]** **,** **2ⁿᵈ** **[L2]** **ENTER**.
The sample statistics will appear on the screen. The sample standard deviation of
the frequency distribution described in columns **L1** and **L2** is 9.2369. In this
example you do not have the actual data. What you have is the frequency
distribution of the data summarized into categories. The standard deviation of this
frequency distribution is an approximation of the standard deviation of the actual
data.

```
1-Var Stats
 x̄=23.2
 Σx=928
 Σx²=24857.1
 Sx=9.23691035
 σx=9.120718173
↓n=40
```

To calculate the sample variance, $(Sx)^2$, type in the value of the standard

deviation at the bottom of the screen and press the $\boxed{x^2}$ key and **ENTER**. The
approximate sample variance is 85.32.

```
 Σx=928
 Σx²=24857.1
 Sx=9.23691035
 σx=9.120718173
↓n=40
9.2369²
        85.32032161
█
```

> ▶ Problem 3 (pg. 155)

(a.) Press **STAT** and select **1:Edit**. Clear **L1** and **L2**. Enter the midpoints for each of the temperature ranges into **L1** and the frequencies ('days') into **L2**. Press **STAT**, highlight **CALC**, select **1:1-Var Stats**, and press 2nd **[L1]** , 2nd **[L2]** **ENTER**.

```
L1      L2      L3      2
54.5    1       ------
64.5    308
74.5    1519
84.5    1626
94.5    503
104.5   11

L2(7) =
```

The population statistics will appear on the screen.

```
1-Var Stats
 x̄=80.43497984
 Σx=319166
 Σx²=25937242
 Sx=8.175217933
 σx=8.174187725
↓n=3968
```

The approximate value of the population mean is 80.43 and the approximate value of the population standard deviation (σx) is 8.174.

(b.) To set up the histogram, push 2nd **[STAT PLOT]** and **ENTER** to select **Plot 1**. Turn ON **Plot 1**, set **Type** to **Histogram**, set **Xlist** to **L1**, set **Freq** to **L2**.

Press **Window** to adjust the Graph Window. Set **Xmin** equal to 54.5 (the midpoint of the first class) and **Xmax** equal to 114.5 (a value that would be the midpoint of an additional class at the end of the table. This extra value is needed to complete the last bar of the histogram). Set **Xscl** equal to 10, which is the class width. Set **Ymin = -5** and **Ymax = 1630**. You do not need to change **Yscl** or **Xres**. Press **GRAPH** and the histogram should appear.

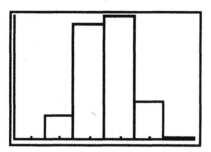

It may look as if there are only four bars in the histogram. There are actually 6 bars. The first and last bars have such small frequencies compared to the other bars that they are extremely small in the graph. Notice that the histogram is bell-shaped.

(c.) The Empirical Rule states that 95% of the data falls in the interval $(\mu \pm 2\sigma)$. To calculate the upper and lower limits of this interval, press **CLEAR** a few times until you get a blank screen. Enter **80.4-2*8.2** to get the lower limit. Press **2ⁿᵈ ENTER** and the calculation will appear again on the screen with the cursor flashing. Move the cursor so that it is positioned on the '-' sign and type in a '+' sign and press **ENTER** to get the upper limit.

Section 3.4

▶ Exploration (pg. 159) Z-scores

Press **STAT** and select **1:Edit**. Clear **L1 and L2**. Enter the home run data from the Table 1 on pg. 114 into **L1**. Press **STAT**, highlight **CALC**, select **1:1-Var Stats**, and press **2ⁿᵈ** [**L1**]**ENTER** to obtain the population mean and standard deviation.

```
1-Var Stats
 x̄=179
 Σx=2506
 Σx²=465278
 Sx=35.8458237
 σx=34.54190002
↓n=14
```

To obtain the Z-scores for the data set, press **STAT** and select **1:Edit.** Highlight **L2** at the top of the second column and press **ENTER**. With the cursor flashing on the bottom line of the screen type in **(L1-179)/34.54.**

```
L1        L2       L3      2
 158      ------   ------
 136
 198
 214
 212
 139
 152
L2 =(L1-179)/34█...
```

The Z-score for each data point in **L1** will appear in **L2**. These Z-scores should have a mean of **0** and a standard deviation of **1**. To check this, press **STAT**, highlight **CALC**, select **1:1-Var Stats**, and press **2ⁿᵈ** [**L2**]**ENTER**

```
1-Var Stats
 x̄=0
 Σx=0
 Σx²=14.0015403
 Sx=1.037806129
 σx=1.000055009
↓n=14
```

Notice that the mean is 0. The population standard deviation is 1.000055009. This value is not *exactly* equal to 1 because we used a rounded value (34.54) for the standard deviation in our calculations.

◀

Problem 9 (pg. 165)

Press **STAT** and select **1:Edit**. Clear **L1** and then enter the data. Press **STAT** and highlight **CALC**. Select **1:1-Var Stats** and press 2nd **[L1]** **ENTER**. The sample mean and sample standard deviation appear on the first screen. Scroll down to the 2nd screen to find the quartiles. **Q1** is the first quartile, **Med** is the 2nd quartile (or median) and **Q3** is the third quartile.

(a.) To calculate the Z-score for the data point 0.97 inches, do the following calculation: (0.97-sample mean)/sample standard deviation.

(c.) The interquartile range(**IQR**) is **Q3-Q1**.

(d.) The lower fence is **Q1-1.5*IQR**. The upper fence is **Q3+1.5*IQR**.

◀

Section 3.5

▶ Example 1 (pg. 169) The Five Number Summary

Press **STAT** and select **1:Edit**. Clear L1 and enter the data from Table 12 on pg. 70 into **L1**. Press **STAT** and highlight **CALC** to display the Calc Menu. Select **1: 1-Var Stats** and press 2nd **[L1]** **ENTER**. Scroll down to the 2nd screen to obtain the five values: **minX, Q1, med, Q3 and maxX.**

```
1-Var Stats
↑n=40
 minX=10.8
 Q₁=17.05
 Med=22.05
 Q₃=28.8
 maxX=47.7
```

◀

▶ Example 2 (pg. 170) A Boxplot

This is a continuation of Example 1 on pg. 169. In Example 1 we entered the data into **L1**.

Press **2ⁿᵈ** [STAT PLOT]. Select **1:Plot 1** and press **ENTER**. Turn On **Plot 1**. Move to the **Type** options. Using the right arrow (you can not use the down arrow to drop to the second line), scroll through the **Type** options and choose the first boxplot which is the first entry in row 2 of the **TYPE** options. Press **ENTER**. Move to **Xlist** and type in **L1**. Press **ENTER** and move to **Freq**. Set **Freq** to **1**. If **Freq** is set on **L2**, press **CLEAR**, and press **ALPHA** to return the cursor to a flashing solid rectangle and type in **1**. Press **ZOOM** and **9** to select **ZoomStat**. The Boxplot will appear on your screen.

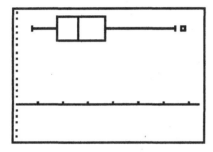

If you press **TRACE** and use the left and right arrow keys, you can display the following information: the smallest data point which is 10.8; Q1 (17.05); the median (22.05); Q3 (28.8); the largest data point which falls inside the upper fence which is 45.9 and the largest data point in the dataset, which is 47.7. This largest value is also an outlier because it lies outside the upper fence. Notice that this value is donated by a small box at the extreme right side of the diagram. (Note: The boxplot does not display the lower and upper fences.)

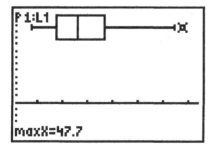

Describing the Relation between Two Variables

CHAPTER

4

Section 4.1

▶ Example 1 (pg. 193)   A Scatter Diagram

Press **STAT**, highlight **1:Edit** and clear **L1** and **L2**. Enter the values of the predictor variable (per capita GDP) into **L1** and the values of the response variable (life expectancy) into **L2**. Press **2ⁿᵈ** **[STAT PLOT]** , select **1:Plot1**, turn **ON** Plot 1 and press **ENTER**. For **Type** of graph, select the **scatter plot** which is the first selection. Press **ENTER**. Enter **L1** for **Xlist** and **L2** for **Ylist**. Highlight the first selection, the small square, for the type of **Mark**. Press **ENTER**. Press **ZOOM** and **9** to select **ZoomStat**.

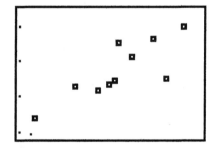

This graph shows a positive linear correlation: as per capita GDP rises, life expectancy also rises.

◀

▶ Example 2 (pg. 197)   The Correlation Coefficient

Press **STAT**, highlight **1:Edit** and clear **L1** and **L2**. Enter the values of the predictor variable (per capita GDP) into **L1** and the values of the response variable (life expectancy) into **L2**. In order to calculate r, the correlation coefficient, you must turn **On** the **Diagnostic** command. Press **2ⁿᵈ [CATALOG]** (Note: CATALOG is found above the **0** key). The CATALOG of functions will appear on the screen. Use the down arrow to scroll to the **DiagnosticOn** command.

Press **ENTER ENTER**.

Press **STAT**, highlight **CALC**, select **4:LinReg(ax+b)** and press **ENTER ENTER**. (Note: This command requires that you specify which lists contain the X-values and Y-values. If you do not specify these lists, the defaults are used. The defaults are: **L1** for the X-values and **L2** for the Y-values.)

```
LinReg
 y=ax+b
 a=.4200227316
 b=68.71511082
 r²=.6551599375
 r=.8094195065
▮
```

The correlation coefficient is r = .8094195065. This suggests a strong positive linear correlation between X and Y.

◀

Section 4.2

▶ Example 2 (pg. 210)   Least Squares Regression Line

Press **STAT**, highlight **1:Edit** and clear **L1** and **L2**. Using Table 3 on pg. 207, enter the values of the predictor variable (per capita GDP) into **L1** and the values of the response variable (life expectancy) into **L2**. Press **STAT**, highlight **CALC** and select **4:LinReg(ax+b)**. This command has several options. One option allows you to store the regression equation into one of the Y-variables. To use this option, with the cursor flashing on the line **LinReg(ax+b)**, press **VARS**.

Highlight **Y-VARS**.

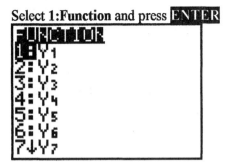

Select **1:Function** and press **ENTER**

Notice that **1:Y1** is highlighted. Press **ENTER**.

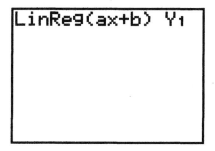

Press **ENTER**.

The output displays the general form of the regression equation: y = ax+b
followed by values for a and b. Next, r^2, the coefficient of determination, and r,
the correlation coefficient , are displayed. If you put the values of a and b into
the general equation, you obtain the specific linear equation for this data:
y= .42x + 68.7151. Press **Y=** and see that this specific equation has been pasted
to **Y1**. (Note: The numerical value for 'b' varies slightly from the book's value
on pg. 211. This difference is simply due to rounding.)

Press **2ⁿᵈ STAT PLOT]** , select **1:Plot1**, turn **ON** Plot1, select **scatter plot**, set
Xlist to **L1** and **Ylist** to **L2**. Press **ZOOM** and **9**.

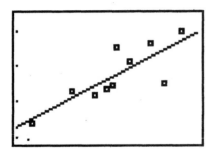

This picture displays a scatter plot of the data and the regression line. The picture indicates a strong positive linear correlation between X and Y, which is confirmed by the r-value of .809.

You can use the regression equation stored in **Y1** to predict Y-values for specific X-values. For example, suppose you would like to use the regression equation to predict the life expectancy for a resident of Italy where the GDP is $21.5. In other words, for X = 21.5, what does the regression equation predict for Y? To find this value for Y, press **VARS**, highlight **Y-VARS**, select **1:Function**, press **ENTER**, select **1:Y1** and press **ENTER**. Press **(** 21.5 **)** and press **ENTER** .

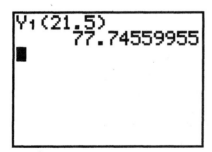

The output shows the predicted Y-value of 77.75 for the input X-value of 21.5.

The residual for Italy is: the actual Y-value for Italy (78.51) – the predicted Y-value for Italy (77.75).

▶ Problem 11 (pg. 217)   The Equation of the Regression Line

Enter the predictor variable values into **L1** and the response variable values into
L2. Press **STAT**. Highlight **CALC**, select **4:LinReg(ax+b)**, press **ENTER**.
Press **VARS**, highlight **Y-VARS**, select **1:Function**, press **ENTER** and select
1:Y1 and press **ENTER**.

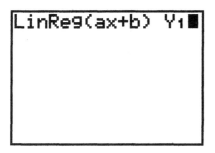

Press **ENTER**.

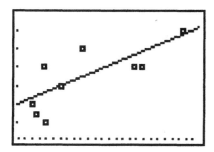

(a.) Using **a** and **b** from the output display, the resulting regression equation is y
= .0261x + 7.8738. Press **Y=** to confirm that the regression equation has been
stored in **Y1**. Press **ZOOM** and **9** for **ZoomStat** and a graph of the scatter plot
with the regression line will be displayed.

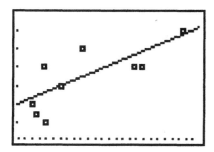

(c.) Next, you can use the regression equation to predict life expectancy for various animals. For example, to predict the life expectancy for an animal with a gestation period of 95 days, press **VARS**, highlight **Y-VARS**, select **1:Function** and press **ENTER**. Select **1:Y1** and press **ENTER**. Press **(** 95 **)** and **ENTER**.

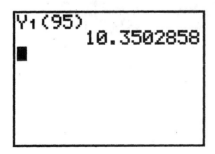

The predicted life expectancy for this species (with a gestation period of 95 days) is 10.35 years.

(d.) To predict the life expectancy of a parakeet, press **2nd** [ENTRY], (found above the ENTER key). Move the cursor so that it is flashing on '9' in the number '95' and type in **18**. Press **ENTER**.

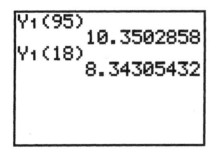

The predicted life expectancy of a parakeet is 8.34 years.

(e.) Press **2nd** [ENTRY] . Move the cursor so that it is flashing on '1' in the number '18' and type in **68**. Press **ENTER**.

The predicted life expectancy of a guinea pig is 9.64 years.

(f.) Residual = Actual Life Expectancy of a guinea pig – predicted life expectancy of a guinea pig: 3 – 9.64 = -6.56.

Section 4.3

▶ Example 1 (pg. 223)   Coefficient of Determination, R^2

Using Table 1 on pg. 193, enter the X-values into **L1** and the Y-values into **L2**.
Press **STAT**. Highlight **CALC**, select **4:LinReg(ax+b)**, press **ENTER**. (For this example, we are not storing the regression equation in **Y1**.) .

```
LinReg
 y=ax+b
 a=.4200227316
 b=68.71511082
 r²=.6551599375
 r=.8094195065
■
```

The value of r^2, .655, is displayed in the output.

◀

▶ Example 2 (pg. 226)   Is a Linear Model Appropriate?

Using the data in Table 6, enter the X-values into **L1** and the Y-values into **L2**. (Note: You may notice that the values displayed in L2 are rounded values. These rounded values are for display purposes only; the actual values are still stored in the calculator.) Press STAT, highlight **CALC** and select **4:LinReg(ax+b)** and press ENTER.

```
LinReg
 y=ax+b
 a=6.755020238
 b=-13469.60059
 r²=.9220401544
 r=.9602292197
```

To plot the residuals, first make sure that there is nothing stored in the Y-registers. Press Y= and check the Y-registers. If any of them contain a function, move the cursor to that Y-register and press CLEAR.

Press 2nd **[STAT PLOT]** , select **1:Plot1**, turn **ON** Plot 1 and press ENTER. For **Type** of graph, select the **scatter plot** which is the first selection. Press ENTER. Enter **L1** for **Xlist**. Move the cursor to **Ylist**. Press 2nd **[List]** and select **7:Resid**. Highlight the first selection, the small square, for the type of **Mark**. Press ENTER. Press ZOOM and 9 to select **ZoomStat**.

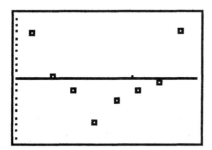

This graph of the residuals vs. the predictor variable (years) shows a U-shaped pattern, which indicates that the linear model is not appropriate.

▶ Example 5 (pg. 228)   Graphical Residual Analysis

Using the data in Table 4 on pg. 215, enter the X-values, (GDP), into **L1** and the Y-values, (life expectancy), into **L2**. Press **STAT**, highlight **CALC** and select **4:LinReg(ax+b)** and press **ENTER**.

To plot the residuals, first make sure that there is nothing stored in the Y-registers. Press **Y=** and check the Y-registers. If any of them contain a function, move the cursor to that Y-register and press **CLEAR**.

Press **2nd [STAT PLOT]** , select **1:Plot1**, turn **ON** Plot 1 and press **ENTER**. For **Type** of graph, select the **scatter plot** which is the first selection. Press **ENTER**. Enter **L1** for **Xlist**. Move the cursor to **Ylist**. Press **2nd [List]** and select **7:Resid**. Highlight the first selection, the small square, for the type of **Mark**. Press **ENTER**. Press **ZOOM** and **9** to select **ZoomStat**.

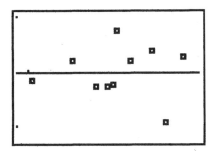

There is no discernable pattern in the plot of the residuals. This indicates support for the linear model that we used to calculate the regression equation.

The next step in analyzing the residuals is to construct a boxplot to determine if there are any unusual residuals values (called outliers.)

Press **2nd [STAT PLOT]** , select **1:Plot1**, turn **ON** Plot 1 and press **ENTER**. For **Type** of graph, select the **boxplot** with outliers, which is the first selection in the second row. Press **ENTER**. Enter **L1** for **Xlist**. Move the cursor to **Freq** and set this equal to **1**. Highlight the first selection, the small square, for the type of **Mark**. Press **ZOOM** and **9** to select **ZoomStat**.

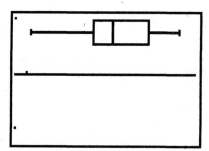

In this example, there are no outliers. This further supports the use of the linear model that was selected for this dataset.

▶ Problem 23 (pg. 234)

Enter the X-values (distance from the sun) into **L1** and the Y-values (sidereal year) into **L2**.

a.) First make sure that there is nothing stored in the Y-registers. Press Y= and check the Y-registers. If any of them contain a function, move the cursor to that Y-register and press CLEAR.

Press **2ⁿᵈ** **[STAT PLOT]** , select **1:Plot1**, turn **ON** Plot 1 and press ENTER. For **Type** of graph, select the **scatter plot** which is the first selection. Press ENTER. Enter **L1** for **Xlist** and **L2** for **Ylist**. Highlight the first selection, the small square, for the type of **Mark**. Press ENTER. Press ZOOM and 9 to select **ZoomStat**.

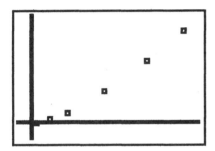

(Note: It is difficult to see all the nine points on this graph. It looks as if there are only 6 data points. In fact, there are nine points but, the first three points are so close together that they are indistinguishable from one another.)

b.) Press STAT, highlight **CALC** and select **4:LinReg(ax+b)** and press ENTER

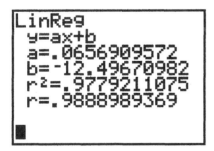

```
LinReg
 y=ax+b
 a=.0656909572
 b=-12.49670982
 r²=.9779211075
 r=.9888989369
```

c.) Press **2ⁿᵈ** **[STAT PLOT]** , select **1:Plot1**, turn **ON** Plot 1 and press ENTER. For **Type** of graph, select the **scatter plot** which is the first selection. Press ENTER. Enter **L1** for **Xlist.** Move the cursor to **Ylist**. Press **2ⁿᵈ** **[List]** and select **7:Resid**. Highlight the first selection, the small square, for the type of **Mark**. Press ENTER. Press ZOOM and 9 to select **ZoomStat**.

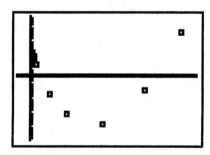

d.) This graph of the residuals vs. the x- variable shows a U-shaped pattern, which indicates that the linear model is not appropriate.

▶ Problem 27 (pg. 234)

Enter the X-values (heights) into **L1** and the Y-values (weights) into **L2**.

a.) First make sure that there is nothing stored in the Y-registers. Press Y= and check the Y-registers. If any of them contain a function, move the cursor to that Y-register and press CLEAR.

Press **2ⁿᵈ [STAT PLOT]** , select **1:Plot1**, turn **ON** Plot 1 and press ENTER. For **Type** of graph, select the **scatter plot** which is the first selection. Press ENTER. Enter L1 for **Xlist** and L2 for **Ylist**. Highlight the first selection, the small square, for the type of **Mark**. Press ENTER. Press ZOOM and 9 to select **ZoomStat**.

b.) Press STAT, highlight **CALC** and select **4:LinReg(ax+b)** and press ENTER

```
LinReg
 y=ax+b
 a=3.283830673
 b=-36.76891048
 r²=.3268359151
 r=.571695649
```

c.) Press STAT and **Edit**. Move the cursor so that it is flashing on Randy Johnson's height of '82' in **L1** and press DEL. Move the cursor so that it is flashing on Randy Johnson's weight of '230' in **L2** and press DEL. Press STAT, highlight **CALC** and select **4:LinReg(ax+b)** and press ENTER

```
LinReg
 y=ax+b
 a=3.948356808
 b=-85.29107981
 r²=.1840465622
 r=.4290064827
```

Section 4.4

▶ Example 4 (pg. 239)   An Exponential Model

The TI-83 has the capability of creating an exponential model directly from the data. It is not necessary to linearize the equation by taking the log of the x-values and y-values.

a.) Enter the X-values into **L1** and the Y-values into **L2**. Make sure that there is nothing stored in the Y-registers. Press Y= and check the Y-registers. If any of them contain a function, move the cursor to that Y-register and press CLEAR. Press **2ⁿᵈ** [STAT PLOT] , select **1:Plot1**, turn **ON** Plot 1 and press ENTER. For **Type** of graph, select the **scatter plot** which is the first selection. Press ENTER. Enter **L1** for **Xlist** and **L2** for **Ylist**. Highlight the first selection, the small square, for the type of **Mark**. Press ENTER. Press ZOOM and 9 to select **ZoomStat**.

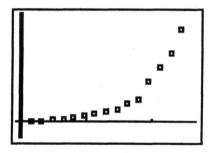

The shape of the scatter plot suggests that an exponential model would be appropriate for this dataset.

(d.) Press STAT, highlight **CALC** and select **0:ExpReg**. Press VARS, highlight **Y-VARS**, select **1:Function**, press ENTER and select **1:Y1** and press ENTER ENTER.

```
ExpReg
 y=a*b^x
 a=.4025702175
 b=1.397449221
 r²=.9812939004
 r=.9906027965
■
```

The exponential equation for this dataset is: $y = .4026(1.3974)^x$.
If you press GRAPH , you can see a picture of the data along with the
exponential model of best fit.

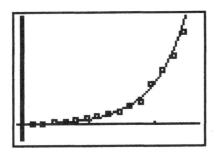

e.) To predict the closing price of Harley Davidson stock at the end of the year
2002, press VARS, highlight **Y-VARS**, select **1:Function** and press ENTER.
Select **1:Y1** and press ENTER. Press (16) and ENTER. (Note: The year 2002
is represented by x =16.)

```
Y₁(16)
         85.15658133
```

(Note: The value in your textbook, $85.11, is slightly different because rounded
values for **a** and **b** were used in the calculation. The TI-83 does not use rounded
values.)

> ▶ Example 5 (pg. 242)   A Power Model

The TI-83 has the capability of creating a power model directly from the data. It is not necessary to linearize the equation by taking the log of the x-values and y-values. (Note: This model can used only in situations where x > 0, y > 0.)

a.) Enter the X-values into **L1** and the Y-values into **L2**. Make sure that there is nothing stored in the Y-registers. Press ▊Y=▊ and check the Y-registers. If any of them contain a function, move the cursor to that Y-register and press ▊CLEAR▊. Press **2ⁿᵈ** ▊[STAT PLOT]▊ , select **1:Plot1**, turn **ON** Plot 1 and press ▊ENTER▊. For **Type** of graph, select the **scatter plot** which is the first selection. Press ▊ENTER▊. Enter **L1** for **Xlist** and **L2** for **Ylist**. Highlight the first selection, the small square, for the type of **Mark**. Press ▊ENTER▊. Press ▊ZOOM▊ and ▊9▊ to select **ZoomStat.**

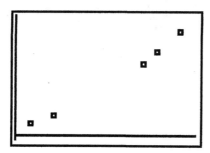

This is a very small dataset and it is difficult to determine which model would be the most appropriate one. For the purposes of this example, we will select the **power model**.

(d.) Press ▊STAT▊, highlight **CALC** and scroll down to **A:PwrReg** and press ▊ENTER▊ . Press ▊VARS▊, highlight **Y-VARS**, select **1:Function**, press ▊ENTER▊ and select **1:Y1** and press ▊ENTER▊.

```
PwrReg
 y=a*x^b
 a=4.933897526
 b=1.992836932
 r²=.9999944648
 r=.9999972324
```

The power equation for this dataset is: $y = 4.934(x)^{1.99284}$

If you press **GRAPH**, you can see a picture of the data along with the power model of best fit.

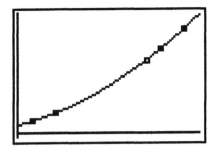

e.) To predict the distance a ball would have fallen if it took 4.2 seconds to hit the ground, press **VARS**, highlight **Y-VARS**, select **1:Function** and press **ENTER**. Select **1:Y1** and press **ENTER**. Press **(4.2)** and **ENTER**.

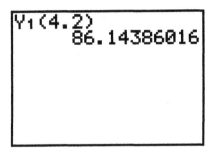

Probability

CHAPTER

5

Section 5.1

▶ Example 6 (pg. 269)   Simulating Probabilities

In this example, we will use simulation to estimate the probability of "having a boy." We assume the simple events, "having a boy," and "having a girl," are equally likely. In this simulation, we will designate "0" as a "boy," and "1" as a "girl."

a.) The first step is to set the *seed* by selecting a 'starting number' and storing this number in **rand**. Suppose, for this example, that we select the number '1204' as the starting number. Type **1204** into your calculator and press the STO key. Next press the MATH key, move the cursor to highlight **PRB**. Select **rand,** and press ENTER. The starting value of '1204' will be stored into **rand** and will be used as the *seed* for generating random numbers. (Note: This process of setting the *seed* is optional. You can omit it and simply go directly to the next step.)

(a.) For 100 births, press MATH, highlight **PRB**, and select **5:randInt(** and press ENTER. The **randInt(** command requires a minimum value, (which is 0 for this simulation), a maximum value (which is 1), and the number of trials (100). In the **randInt(** command type in **0** , **1** , **100.**

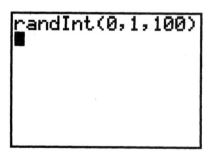

Press ENTER. It will take a few seconds for the calculator to generate 100 numbers. Notice, in the upper right hand corner a flashing | |, indicating that the calculator is working. When the simulation has been completed, a string of **0's**

and **1's** will appear on the screen followed by **….**, indicating that there are more numbers in the string that are not shown.

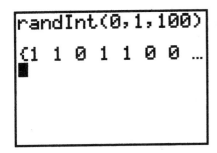

Press **STO** and 2nd **[L1]** **ENTER**. This will store the string of numbers in **L1**. Press 2nd **[LIST]** and highlight **MATH**. Select **sum(** and type in **L1**.

```
randInt(0,1,100)

{1 1 0 1 1 0 0 …
Ans→L1
{1 1 0 1 1 0 0 …
sum(L1)
                 62
```

The sum of **L1** equals the number of "1's" in the list. Since we have designated a "1" to be a "girl", we have 62 girls in the simulation. So, based on this simulation, we *estimate* the probability that a child is born a "girl" to be 62 out of 100 or 62%. And, the *estimated* probability that a child is a born a "boy" is 38% (100%-62%). These values are quite different from the theoretical probabilities. (The theoretical probabilities of both of these events are 50%). This difference is due to the fact that we did not do a very large number of repetitions.

(b.) Repeat the steps in part (a.) and increase the sample size to 999. This is the maximum sample size that the calculator will allow.

◀

▶ Problem 31   (pg.274)

In this simulation, we will use the integers 1,2,3,4,5 and 6 to represent the six possible outcomes on a six-sided die. Press MATH, highlight **PRB**, and select **5:randInt(** and press ENTER. The **randInt(** command requires a minimum value, (which is 1 for this simulation), a maximum value (which is 6), and the number of trials (100). In the **randInt(** command type in **1** ⌐,⌐ **6** ⌐,⌐ **100.**

```
randInt(1,6,100)
```

Press ENTER. It will take several seconds for the calculator to generate 100 rolls of the die. Notice, in the upper right hand corner a flashing ⌐|⌐, indicating that the calculator is working. When the simulation has been completed, a string of **0's, 1's, 2's. etc.** will appear on the screen followed by ...., indicating that there are more numbers in the string that are not shown.

Store the data in **L1** by pressing STO and 2nd **[L1]** ENTER.

(a.) One way to count the number of 1's in your simulation is to create a histogram of the results. First make sure that there is nothing stored in the Y-registers. Press Y= and check the Y-registers. If any register contains a function, move the cursor to that Y-register and press CLEAR.

To set up the histogram, push 2nd **[STAT PLOT]** and ENTER to select **Plot 1.** Turn **ON Plot 1,** set **Type** to **Histogram,** set **Xlist** to L1,. set **Freq** to 1. Press Window to adjust the Graph Window. Set **Xmin** equal to 1 (the minimum value in your simulation) and **Xmax** equal to 7 (a value that would be one integer larger than the maximum value on the roll of a die. This extra value is needed to complete the last bar of the histogram). Set **Xscl** equal to 1. (Note: In many cases it is not necessary to change the values for **Ymin, Ymax** or **Yscl.** What you must do is to check these values and make sure that **Ymin** is a small negative value (between –5 and –1 would be good) and **Ymax** must be larger than the largest frequency value in your dataset. A good value for **Ymax** would be 30. You never need to adjust **Yscl.**

Press GRAPH and the histogram should appear.

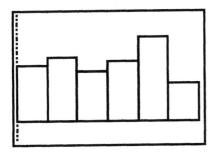

You can press ████ and scroll through the bars of the histogram.
The first bar of the histogram represents all the rolls that resulted in 1's.

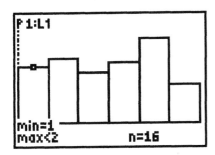

Notice, for this example, there were 16 rolls of 1's. So, based on this simulation,
the estimated probability of rolling a '1' is 16 out of 100 or 16%.

(b.) Repeat the simulation.

(c.) Repeat the simulation and increase the number of rolls of the die to 500.

◀

Section 5.5

▶ Example 4 (pg. 304)   The Traveling Salesman - Factorials

The total number of different routes that are possible can be computed using the factorial function.

Press $\boxed{7}$, MATH, highlight **PRB** and select **4:!** and ENTER.

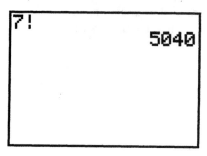

There are 5,040 different possible routes.

◀

▶ Example 5 (pg. 305) Permutations

(a.) In this example, there are 7 objects (n=7). From these 7 objects, 5 objects are selected (r=5). The permutation formula counts the number of different ways that these 5 objects can be selected and arranged from the total of 7 objects. The formula **nPr** is used with **n = 7** and **r =5.** So, the formula is **7P5**.

Press ▮ , MATH, highlight **PRB** and select **2:nPr** and ENTER.

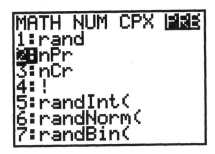

Now press ▮ and ENTER. The answer, 2520, appears on the screen.

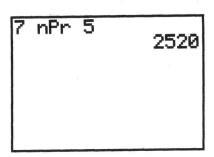

▸ Example 8 (pg. 307) Combinations

(b.) In this example, there are 6 objects (n=6). From these 6 objects, 4 objects are selected (r=4). The combination formula counts the number of different ways that these 4 objects can be selected from the total of 6 objects. The formula **nCr** is used with **n =6** and **r =4.** So, the formula is **6C4**.

Press **6**, **MATH**, highlight **PRB** and select **3:nCr** and **ENTER**.

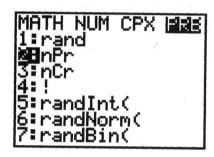

Now press **4**, and **ENTER**. The answer, 15, appears on the screen.

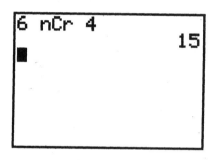

▶ Example 11 (pg. 309) Arranging Flags - Permutations with Repetitions

To calculate $\dfrac{10!}{5!3!2!}$ you will use the factorial function (!). Enter the first value, 10, press MATH, highlight **PRB** and select **4:!** . Then press ÷ . Open the parentheses by pressing (. Enter the next value, **5**, press MATH, **PRB**, and select **4:!** . To multiply by 3!, press x and enter the next value, **3**. Press MATH, **PRB**, and select **4:!** . To multiply by 2!, press x and enter the next value, **2**. Press MATH, **PRB**, and select **4:!** . Close the parentheses) and press ENTER.

```
10!/(5!*3!*2!)
              2520
```

◀

> ▶ **Example 12 (pg. 310)**   Winning the Lottery - Probabilities
> involving Combinations

To calculate the probability of winning the Illinois Lottery, you must calculate

$$\frac{2}{_{54}C_6}.$$

Enter the numerator, 2, into your calculator. Next press █ and enter the first value in the denominator, **54**, press MATH, highlight **PRB** and select **3:nCr**, enter the next value, **6**. Press ENTER and the answer will be displayed on your screen.

```
2/54 nCr 6
      7.7437845E-8
```

Notice that the answer appears in scientific notation. To convert to standard notation, move the decimal point **8** places to the left. The answer is .0000000774.

◀

▶ Example 13 (pg. 310) Probabilities involving Combinations

In this example, there are 120 fasteners in the shipment. Four fasteners in the shipment are defective. The remaining 116 fasteners are not defective. The quality-control manager randomly selects five fasteners.

To calculate the probability of selecting exactly one defective fastener, you must calculate: $\dfrac{{}_4C_1 * {}_{116}C_4}{{}_{120}C_5}$

To calculate the numerator, enter the first value, **4**, press MATH, highlight **PRB** and select **3:nCr** and enter the next value, **1**. Next press × and enter the next value, **116**, press MATH, highlight **PRB** and select **3:nCr**, enter the next value, **4**. Next press ÷ and enter the first value in the denominator, **120**, press MATH, highlight **PRB** and select **3:nCr**, enter the next value, **5**. Press ENTER and the answer will be displayed on your screen.

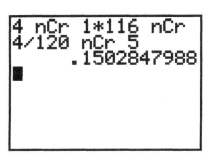

```
4 nCr 1*116 nCr
4/120 nCr 5
       .1502847988
■
```

▶ Problem 57 (pg. 313)

In this exercise, there are two groups made up of 8 students and 10 faculty. The combined number in the two groups is 18. Five individuals are to be selected from the total.

(a.) To select all students, you must choose 5 students from the group of 8 students and 0 faculty from the group of 10. Using the combination formula, you will do the following calculation: $\dfrac{_8C_5 \, * \, _{10}C_0}{_{18}C_5}$.

To calculate the numerator, enter the first value, **8**, press **MATH**, highlight **PRB** and select **3:nCr** and enter the next value, **5**. Next press **x** and enter the next value, **10**, press **MATH**, highlight **PRB** and select **3:nCr**, enter the next value, **0**. Next press **÷** and enter the first value in the denominator, **18**, press **MATH**, highlight **PRB** and select **3:nCr**, enter the next value, **5**. Press **ENTER** and the answer will be displayed on your screen.

```
8 nCr 5*10 nCr 0
/18 nCr 5
          .0065359477
■
```

(b.) Repeat the steps in part (a.) but select 0 students and 5 faculty.

(c.) Repeat the steps in part (a.) but select 2 students and 3 faculty.

▶ Problem 61 (pg. 313)

The compact disk has a total of 13 songs. Of the 13 songs, there are 5 songs that you like and 8 songs that you do not like. Suppose that four songs are randomly selected and played.

(a.) Calculate the probability that, among these first four songs selected, you like exactly two of them. Using the combination formula, you will do the following calculation: $\dfrac{{}_5C_2 * {}_8C_2}{{}_{13}C_4}$.

To calculate the numerator, enter the first value, **5**, press MATH, highlight **PRB** and select **3:nCr** and enter the next value, **2**. Next press x and enter the next value, **8**, press MATH, highlight **PRB** and select **3:nCr**, enter the next value, **2**. Next press ÷ and enter the first value in the denominator, **13**, press MATH, highlight **PRB** and select **3:nCr**, enter the next value, **4**. Press ENTER and the answer will be displayed on your screen.

(b.) Repeat the steps in part (a.) but select 3 songs from the group of songs that you like and 1 from the group of songs that you do not like.

(c.) Repeat the steps in part (a.) but select 4 songs from the group of songs that you like and 0 songs from the group that you do not like.

◀

Discrete Probability Distributions

Section 6.1

▶ Example 4 (pg. 328) A Probability Histogram

Press STAT and select **1:EDIT**. Clear **L1** and **L2**. Using Table 1 on pg. 327, enter the X-values into **L1** and the P(X=x) values into **L2**.

To graph the probability distribution, press 2nd [STAT PLOT] and press ENTER. Turn **ON** Plot 1, select **Histogram** for **Type**, type in 2nd [L1] for **Xlist** and 2nd [L2] for **Freq.** Press WINDOW and set **Xmin** = 0, **Xmax** = 4, **Xscl** = 1, **Ymin** = **0** and **Ymax** = .55. Choosing 'Xmax=4' leaves some space at the right of the graph in order to complete the histogram. The Ymax value was selected by looking through the values in **L2** and then rounding the largest value UP to a convenient number. Press GRAPH to view the histogram.

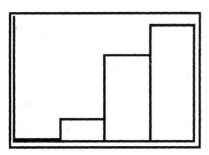

◀

▶ Example 5 (pg. 330) The Mean of a Probability Distribution

Press STAT and select **1:EDIT**. Clear **L1** and **L2**. Enter the X-values from Table 3 into **L1** and the P(X=x) values into **L2**. Press STAT and highlight **CALC**. Select **1:1-Var Stats,** press ENTER and press 2nd [L1] [,] 2nd [L2] ENTER to see the descriptive statistics.

```
1-Var Stats L₁,L
 2




```

```
1-Var Stats
 x̄=2.4
 Σx=2.4
 Σx²=6.24
 Sx=
 σx=.692820323
↓n=1
```

The mean of this discrete random variable is 2.4.

◀

> ▶ Example 7 (pg. 332) The Variance and Standard Deviation of a
> Discrete Random Variable

Using the data from Example 5 on pg. 330, which you stored in **L1** and **L2**, press
STAT highlight **CALC**. Select **1:1-Var Stats,** press **ENTER** and press 2nd **[L1]**
, 2nd **[L2]** **ENTER**. The population standard deviation of the discrete random
variable , σx, is .6928.

```
1-Var Stats
 x̄=2.4
 Σx=2.4
 Σx²=6.24
 Sx=
 σx=.692820323
↓n=1
■
```

To find the variance, you must use the value of the standard deviation. Since the
variance is equal to the standard deviation squared, type in **.69282** and press the
x^2 key. The variance of the discrete random variable is **.4799955**.

```
 Σx=2.4
 Σx²=6.24
 Sx=
 σx=.692820323
↓n=1
.69282²
        .4799995524
■
```

▶ Example 8 (pg. 334)   The Expected Value

Press **STAT** and select **1:EDIT**. Clear **L1** and **L2**. Enter the X-values from Table 6 into **L1** and the associated probabilities into **L2**.

```
L1        L2        L3       3
 350      .99865   ------
-2.5E5    .00135
------    ------


L3(1)=
```

Notice that the value of –249,650 appears as –2.5E5. This is a rounded value and it is written in scientific notation. The actual value is stored in the calculator; the rounded value is for display purposes only.

Press **STAT** and highlight **CALC**. Select **1:1-Var Stats,** press **ENTER** and press 2nd **[L1]** **,** 2nd **[L2]** **ENTER** to see the descriptive statistics.

```
1-Var Stats
 x̄=12.5
 Σx=12.5
 Σx²=84261250
 Sx=
 σx=9179.384171
↓n=1

```

The expected value of this discrete random variable is 12.5.

◀

▶ Problem 17 (pg. 337)

(a.) Enter the X-values into L1 and the frequencies into L2. Press 2nd [QUIT] .
Press 2nd [LIST] and select **MATH**. Select **5:sum(** and type in **L2**. Press
ENTER. The answer is the sum of the frequencies in L2. Press STAT and select
1:EDIT. Move the cursor to highlight 'L2' at the top of the second list and press
ENTER . With the cursor flashing at the bottom of the screen type in **L2** ÷ (the
sum of L2). This will convert the frequencies in L2 into probabilities. To confirm
that you now have a probability distribution represented in L1 and L2, press 2nd
[QUIT] . Press 2nd [LIST] and select **MATH**. Select **5:sum(** and type in **L2**.
Press ENTER. The answer is the sum of the probabilities in L2. This sum should
equal 1.

(b.) To draw the probability histogram, press 2nd [STAT PLOT] and press
ENTER. Turn **ON** Plot 1, select **Histogram** for **Type**, type in 2nd [L1] for **Xlist**
and 2nd [L2] for **Freq**. Press WINDOW and set **Xmin = 1, Xmax = 9, Xscl = 1,
Ymin = 0** and **Ymax = .14.** Choosing 'Xmax=9' leaves some space at the right of
the graph in order to complete the histogram. The Ymax value was selected by
looking through the values in **L2** and then rounding the largest value UP to a
convenient number. Press GRAPH to view the histogram.

(c.-d.) Press STAT and highlight **CALC**. Select **1:1-Var Stats,** press ENTER and
press 2nd [L1] , 2nd [L2] ENTER to see the descriptive statistics.

Section 6.2

▶ Example 3 (pg. 344) The Binomial Probability Distribution

(a.) To find the probability that exactly 10 households have cable, we will use the binomial probability density function, **binompdf(n,p,x).** For this example, n = 15, p = .75 and x = 10. Press 2ⁿᵈ [DISTR]. Scroll down through the menu to select **0:binompdf(** and press ENTER . Type in **15** ⎡,⎤ **.75** ⎡,⎤ **10**) and press ENTER. The answer, **.1651,** will appear on the screen.

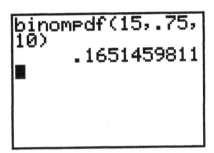

```
binompdf(15,.75,
10)
        .1651459811
■
```

b-c. To calculate inequalities, such as the probability that *at least* 13 households have cable, P(X ≥ 13) or the probability that *fewer* than 13 households have cable, P(X < 13) , you can use the cumulative probability command: **binomcdf (n,p,x).** This command accumulates probability starting at X = 0 and ending at a specified X-value.

To calculate P(X ≥ 13), press 2ⁿᵈ [DISTR] and select **A:binomcdf(** by scrolling through the options and selecting **A:binomcdf(** or by pressing ALPHA A. (Note: A is the ALPHA function on the MATH key.) Type in **15** ⎡,⎤ **.75** ⎡,⎤ **12** ⎡)⎤ and press ENTER. The result, P(X ≤ 12) = .764. This value is the *complement* of P(X ≥ 13). Subtract this value from 1 to obtain P(X ≥ 13).

P(X < 13) is the same as P(X ≤ 12).

◀

▶ Example 5 (pg. 346) Binomial Probability Histograms

a.) Construct a probability distribution for a binomial probability model with n = 10 and p = 0.2. Press STAT, select **1:EDIT** and clear **L1** and **L2**. Enter the values 0 through 10 into **L1**. Press 2nd [QUIT].

To calculate the probabilities for each X-value in **L1**, first change the display mode so that the probabilities displayed will be rounded to 3 decimal places. Press MODE and change from **FLOAT** to **3**. Press ENTER. This will round each of the probabilities to 3 decimal places. Press 2nd [QUIT].

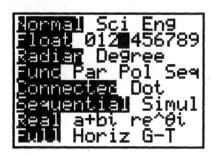

Next press 2nd [DISTR] and select **0:binompdf(** and type in **10** , **.2**) and press ENTER. Store these probabilities in **L2** by pressing STO 2nd [L2] ENTER.

```
binompdf(10,.2)
{.107 .268 .302...
Ans→L₂
{.107 .268 .302...
■
```

To graph the binomial distribution, press 2nd [STAT PLOT] and press ENTER. Turn **ON** Plot 1, select **Histogram** for **Type**, type in 2nd [L1] for **Xlist** and 2nd [L2] for **Freq.** Adjust the graph window by pressing WINDOW and setting **Xmin** = 0, **Xmax** = 11, **Xscl** = 1, **Ymin** = 0 and **Ymax** = **.31.** Choosing 'Xmax=11' leaves some space at the right of the graph in order to complete the histogram. The Ymax value was selected by looking through the values in **L2** and then rounding the largest value UP to a convenient number. Press GRAPH to view the histogram.

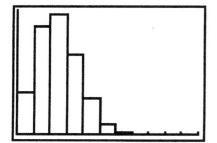

▶ Problem 17 (pg. 352)

(a.) Construct a probability distribution for a binomial probability model with n = 9 and p = .75. Press STAT, select **1:EDIT** and clear **L1** and **L2**. Enter the values 0 through 9 into **L1**. Press 2nd [QUIT]. To calculate the probabilities for each X-value in **L1**, first change the display mode so that the probabilities displayed will be rounded to 3 decimal places. Press MODE and change from **FLOAT** to **3** and press 2nd [QUIT].

Next press 2nd [DISTR] and select **0:binompdf(** and type in **9** , **.75**) and press ENTER. Store these probabilities in **L2** by pressing STO 2nd [L2] .

(b.) Press STAT and highlight **CALC**. Select **1:1-Var Stats,** press ENTER and press 2nd [L1] , 2nd [L2] ENTER to see the descriptive statistics.

(c.) Use the formulas for the mean and standard deviation of a binomial random variable. The mean is: $\mu = n * p$; the standard deviation is: $\sigma = \sqrt{n * p * q}$.

(d.) To graph the binomial distribution, press 2nd [STAT PLOT] and press ENTER. Turn **ON** Plot 1, select **Histogram** for **Type**, type in 2nd [L1] for **Xlist** and 2nd [L2] for **Freq**. Adjust the graph window by pressing WINDOW and setting **Xmin = 0, Xmax = 10, Xscl = 1, Ymin = 0** and **Ymax = .31.** Choosing 'Xmax=10' leaves some space at the right of the graph in order to complete the histogram. The Ymax value was selected by looking through the values in **L2** and then rounding the largest value UP to a convenient number. Press GRAPH to view the histogram.

> ▶ Problem 41 (pg. 355)

(a.) To generate random samples for this binomial model, press **MATH**, select **PRB** and select **7:randBin(**. This command requires three values: **n**, which is the sample size; **p**, the probability; and **x**, the number of samples. For this example type in **30** ⬚ **.98** ⬚ **100)**. Press ENTER It will take the calculator a few minutes to complete this simulation.

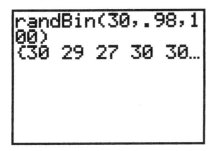

Store these probabilities in **L1** by pressing STO 2ⁿᵈ [L1] .

(b.) To use the results of the simulation to compute the probability that exactly 29 of the 30 males survive to age 30, construct a histogram of the simulation. Press 2ⁿᵈ [STAT PLOT] and press ENTER. Turn **ON** Plot 1, select **Histogram** for **Type**, type in 2ⁿᵈ [L1] for **Xlist** and **1** for **Freq.** Adjust the graph window by pressing WINDOW and setting **Xmin = 25, Xmax = 31, Xscl = 1, Ymin = 0** and **Ymax = 60**. Press GRAPH to view the histogram. Press TRACE and scroll through the bars until you reach the bar for '29'. Take the frequency for that bar and divide it by 100 (the total number of simulations). Your result is the probability that exactly 29 males in a sample of 30 males will survive to age 30.

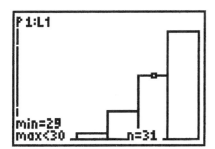

In this simulation, the probability is 31 out of 100 or 31%.

(c.) Press 2ⁿᵈ DISTR and select **0:binompdf** and enter **30** ⬚ **.98** ⬚ **29)**.

(d.) Press GRAPH and the histogram of the simulation will appear. Press TRACE and scroll through the bars for '28', '29' and '30'. Sum the frequencies for these bars. Divide this sum by 100. This value is P(X≥28). The *complement* of this is P(X≤ 27). Subtract P(X≥28) from 1 to get P(X≤ 27).

(e.) Press 2nd DISTR and select **A:binomcdf** and enter **30** , **.98** , **27).** This value is P(X≤ 27).

(f-g.) First, calculate the mean and standard deviation of the 100 simulations. Press STAT and highlight **CALC**. Select **1:1-Var Stats,** press ENTER and press 2nd [L1] ENTER to see the descriptive statistics. Then, use the formulas for the mean and standard deviation of a binomial random variable. The mean is: $\mu = n * p$; the standard deviation is: $\sigma = \sqrt{n * p * q}$.

▶ Problem 44 (pg. 355)

(a.) Suppose the probability that Shaquille O'Neal makes a *free throw* is .536. To find the probability that the *first* free throw he makes occurs on his third shot, press 2ⁿᵈ [DISTR] and select **D:geometpdf(** and type in **.536** ⸴ **3** ⟩ .

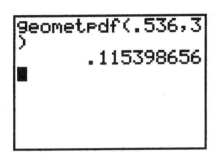

(b.) Construct a probability distribution for a geometric probability model with p = .563. Press STAT, select **1:EDIT** and clear **L1** and **L2**. Enter the values 1 through 10 into **L1**. Press 2ⁿᵈ [QUIT]. To calculate the probabilities for each X-value in **L1**, first change the display mode so that the probabilities displayed will be rounded to 3 decimal places. Press MODE and change from **FLOAT** to **3** and press 2ⁿᵈ [QUIT].

Next press 2ⁿᵈ [DISTR] and select **D:geometpdf(** and type in **.563** ⸴ **L1** ⟩ and press ENTER. Store these probabilities in **L2** by pressing STO 2ⁿᵈ [L2] .

(c.) Press STAT and highlight **CALC**. Select **1:1-Var Stats,** press ENTER and press 2ⁿᵈ [L1] ⸴ 2ⁿᵈ [L2] ENTER to see the descriptive statistics.

(d.) Calculate the mean of a geometric probability model using the formula:

$$\mu = \frac{1}{p} .$$

Section 6.3

▶ **Example 2 (pg. 359)** Probabilities of a Poisson Process

For problems that can be modeled with the poisson probability model, either the values of λ and t are given or the value of μ is given. These parameters (λ, **t and** μ) are related to each other in the following way: $\mu = \lambda * t$. The **poissonpdf** command requires a value for μ.

(a.) In this example, $\lambda = 2$ (cars per minute) and $t = 5$ (minutes) and, therefore, $\mu = 2 * 5 = 10$. Use the command **poissonpdf (μ,x)** with $\mu = 10$ and $X = 6$. Press 2nd [DISTR] and select **B:poissonpdf(** and type in **10** , **6**) and press ENTER. The answer will appear on the screen.

```
poissonpdf(10,6)
        .063055458
```

(b.) To calculate the probability that *less than 6* cars arrive in the 5 minute time period, use the command **poissoncdf (μ,x)** with $\mu = 10$ and $X = 5$. Press 2nd [DISTR] and select **C:poissoncdf(** and type in **10** , **5**) and press ENTER. The answer will appear on the screen.

```
poissoncdf(10,5)
       .0670859629
```

(c.) $P(X \geq 6) = 1 - P(X \leq 5) = 1 - .0671 = .9329$.

▶ Problem 5 (pg. 366)

This is an example of a Poisson process with μ=**5.**

(a.) Press 2nd [DISTR] and select **B:poissonpdf(** and type in 5 ⬚ 6).

(b.) To calculate P(X < 6) press 2nd [DISTR] and select **C:poissoncdf(** and type in 5 ⬚ 5).

(c.) P(X ≥ 6) = 1 – P(X < 6)

(d.) Press 2nd [DISTR] and select **B:poissonpdf(** and type in 5 ⬚ 2nd ⬚ 2 ⬚ **3** ⬚ **4** 2nd ⬚ ⬚ and press ENTER.

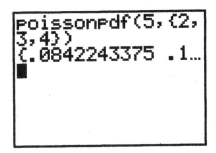

The first answer that appears in the output is P(X=2) which is .0842. Use the right arrow to scroll to the right to see P(X=3) and P(X=4).

The Normal Probability Distribution

CHAPTER

7

Section 7.2

▶ Example 1 (pg. 393) Area Under the Standard Normal Curve
to the left of a Z-score

In this example of the standard normal curve, we will calculate the area to the *left* of Z = 1.68.

The TI-83 has two methods for calculating this area.

Method 1: **Normalcdf(**lowerbound, upperbound, μ, σ) computes the area between a lowerbound and an upperbound. In this example, you are computing the area from *negative infinity* to 1.68. Negative infinity is specified by (-) 1 2nd [EE] 9 9 (Note: **EE** is found above the comma ,). Try entering −1 EE 99 into your calculator.

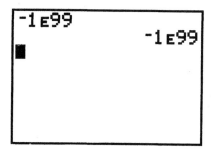

Now, to calculate the area to the left of 1.68, press 2nd [DISTR] and select **2:normalcdf(** and type in -1E99 , 1.68 , 0 , 1) and press ENTER. (Note: For the standard normal curve, $\mu = 0$ and $\sigma = 1$.)

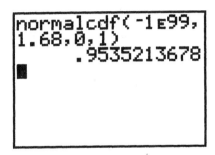

Method 2: This method calculates the area and also displays a graph of the probability distribution. You must first set up the WINDOW so that the graph will be displayed properly. Press **WINDOW** and set **Xmin** equal to -3 and set **Xmax** equal to 3. Set **Xscl** equal to 1.

Setting the Y-range is a little more difficult to do. A good "rule - of - thumb" is to set **Ymax** equal to .5 / σ. For this example, set **Ymax = .5.**

Use the blue up arrow to highlight **Ymin**. A good value for **Ymin** is **(-) Ymax / 4** so type in **(-) .5 / 4.**

Press 2nd [QUIT]. Clear all the previous drawings by pressing 2nd **[DRAW]** and selecting **1:ClrDraw** and pressing **ENTER ENTER**. Press 2nd **[STATPLOT]** and TURN OFF all PLOTS. Now you can draw the probability distribution. Press 2nd [DISTR]. Highlight **DRAW** and select **1:ShadeNorm(** and type in **-1E99 , 1.68 , 0 , 1)** and press **ENTER**. The output displays a normal curve with the appropriate area shaded in and its value computed.

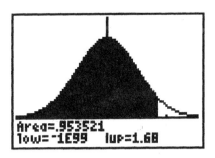

▶ Example 2 (pg. 395) Area Under the Standard Normal Curve
to the right of a Z-score

In this example of the standard normal curve, we will calculate the area to the *right* of Z = -0.46.

Method 1: **Normalcdf(**lowerbound, upperbound, μ, σ) computes the area between a lowerbound and an upperbound. In this example, you are computing the area from –0.46 to *positive infinity*. Positive infinity is specified by 1 2nd [EE] 9 9 (Note: **EE** is found above the comma ,).

To calculate the area to the right of –0.46, press 2nd [DISTR] and select **2:normalcdf(** and type in –0.46 , 1E99 , 0 , 1) and press ENTER.

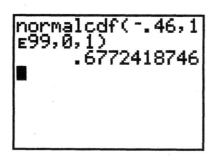

Method 2: This method calculates the area and also displays a graph of the probability distribution. Press WINDOW and set **Xmin** equal to -3 and set **Xmax** equal to 3. Set **Xscl** equal to 1. Set **Ymax = .5**. Set **Ymin = .5/4**. Press 2nd [QUIT]. Clear all the previous drawings by pressing 2nd [DRAW] and selecting **1:ClrDraw** and pressing ENTER ENTER. Press 2nd [STATPLOT] and **TURN OFF** all **PLOTS**. Now you can draw the probability distribution. Press 2nd [DISTR]. Highlight **DRAW** and select **1:ShadeNorm(** and type in **–0.46** , 1E99 , 0 , 1) and press ENTER. The output displays a normal curve with the appropriate area shaded in and its value computed.

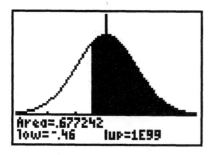

► Example 3 (pg. 396) Area Under the Standard Normal Curve
between two Z-scores

In this example of the standard normal curve, we will calculate the area between
Z= -1.35 and Z= 2.01.

Method 1: **Normalcdf**(lowerbound, upperbound, μ, σ) computes the area
between a lowerbound and an upperbound. In this example, you are computing the
area from −1.35 to 2.01.

To calculate the area between −1.35 and 2.01, press 2nd [DISTR] and select
2:normalcdf(and type in **−1.35** $\boxed{,}$ **2.01** $\boxed{,}$ **0** $\boxed{,}$ **1** $\boxed{)}$ and press ENTER.

```
normalcdf(-1.35,
2.01,0,1)
        .8892764236
```

Method 2: This method calculates the area, and also displays a graph of the
probability distribution. Press WINDOW and set **Xmin** equal to -3 and set **Xmax**
equal to 3. Set **Xscl** equal to 1. Set **Ymax** = .5. Set **Ymin** = .5/4.

Press 2nd [QUIT]. Clear all the previous drawings by pressing 2nd [DRAW]
and selecting **1:ClrDraw** and pressing ENTER ENTER. Press 2nd
[STATPLOT] and **TURN OFF** all **PLOTS**. Now you can draw the probability
distribution. Press 2nd [DISTR]. Highlight **DRAW** and select **1:ShadeNorm(**
and type in **−1.35** $\boxed{,}$ **2.01** $\boxed{,}$ **0** $\boxed{,}$ **1** $\boxed{)}$ and press ENTER. The output displays a
normal curve with the appropriate area shaded in and its value computed.

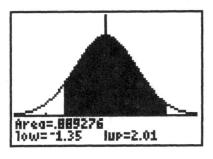

◄

▶ Example 4 (pg. 397) Finding a Z-Score from a Specified
Area to the Left

This is called an inverse normal problem and the command **invNorm(area,** μ, σ) is used. In this type of problem, an area under the normal curve is given and you are asked to find the corresponding Z-score. In this example, the area given is the area to the *left* of a Z-score. The area is 0.32 . (The area value that you enter into the TI-83 must always be area to the left of a Z-score.) .

To find the Z-score corresponding to *left area* of 0.32, press 2nd [DISTR] and select **3:invNorm(** and type in **.32** [,] **0** [,] **1**) and press ENTER.

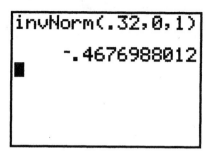

```
invNorm(.32,0,1)
        -.4676988012
```

The Z-score of -.47 has an area of 0.32 to the *left*.

> ▶ Example 5 (pg. 399) Finding a Z-Score from a Specified
>   Area to the Right

This is an inverse normal problem and the command **invNorm(area, μ, σ)** is used. In this type of problem, an area under the normal curve is given and you are asked to find the corresponding Z-score. In this example, the area given is the area to the *right* of a Z-score. The area is 0.4332. (The area value that you enter into the TI-83 must always be area to the left of a Z-score.).

To find the Z-score corresponding to *right area* of 0.4332, subtract 0.4332 from 1 to obtain the area to the *left* of the Z-score. Press 2nd [DISTR] and select **3:invNorm(** and type in **.5668** [,] **0** [,] **1**) and press ENTER.

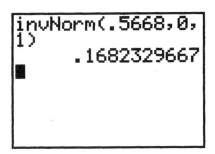

```
invNorm(.5668,0,
1)
         .1682329667
■
```

The Z-score that has an area to the *right* equal to 0.4332 is Z= .17.

◀

▶ Example 6 (pg. 400) Finding Z-Scores for an Area in the Middle

This is an inverse normal problem and the command **invNorm(area,** μ , σ) is used. In this problem the *middle area* is .90. That leaves an area of .10 to be equally divided between the *left* and *right* tail areas. Each of these areas are, therefore, equal to .05. The Z-score that marks the lower edge of the middle area is the Z-score that corresponds to a *left area* of .05.

To find the Z-score corresponding to a *left area* of 0.05, press 2nd [DISTR] and select **3:invNorm(** and type in **.05** [**,**] **0** [**,**] **1**) and press ENTER.

```
invNorm(.05,0,1)
        -1.644853626
```

The Z-score that marks the lower edge of the middle area of .90 is **–1.645.** The Z-score that marks the upper edge is **1.645** (because the normal curve is symmetry about the mean.)

◀

Section 7.3

▶ Example 1 (pg. 406) Finding Area Under a Normal Curve

In this exercise, use a normal distribution with $\mu = 38.72$ and $\sigma = 3.17$.

Method 1:To find the percent of three-year-old females with heights less than 35 inches we calculate P(X < 35). Press 2nd [DISTR] , select **2:normalcdf(** and type in **-1E99** ⌷, **35** ⌷, **38.72** ⌷, **3.17** ⌷ and press ENTER.

```
normalcdf(-1E99,
35,38.72,3.17)
        .1202974198
```

Method 2: To find P(X < 35) and include a graph, you must first set up the **WINDOW** so that the graph will be displayed properly. You will need to set Xmin equal to (μ - 3 σ) and Xmax equal to (μ + 3 σ). Press WINDOW and set **Xmin** equal to (μ - 3 σ) by typing in **38.72 - 3 * 3.17**. Press ENTER and set **Xmax** equal to (μ + 3 σ) by typing in **38.72 + 3 * 3.17**. Set **Xscl** equal to σ, which is 3.17.

Setting the Y-range is a little more difficult to do. A good "rule - of - thumb" is to set **Ymax** equal to .5 / σ. For this example, set **Ymax = .5/3.17**.

Use the blue up arrow to highlight **Ymin**. A good value for **Ymin** is **(-) Ymax / 4** so type in (-) **.158 / 4**.

Press 2nd [DRAW] and select **1:ClrDraw** and press ENTER ENTER. Press 2nd [STATPLOT] and **TURN OFF** all **PLOTS**. Press 2nd [DISTR], highlight **DRAW** and select **1:ShadeNorm(** and type in **-1E99** ⌷, **35** ⌷, **38.72** ⌷, **3.17** ⌷ and press ENTER.

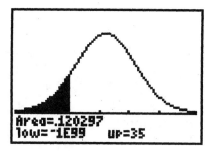

Conclusion: 12.0 % of all three-year-old females are less than 35 inches tall.

Note: When using the TI-83 (or any other technology tool), the answers you obtain may vary slightly from the answers that you would obtain using the standard normal table. Consequently, your answers may not be exactly the same as the answers found in your textbook. The differences are simply due to rounding.

▶ Example 2 (pg. 408) Finding the Probability of a Normal
Random Variable

In this exercise, use a normal distribution with $\mu = 38.72$ and $\sigma = 3.17$.

Method 1: To find P(35 ≤ X ≤ 40) press 2nd [DISTR] , select **2:normalcdf(** and
type in **35** [,] **40** [,] **38.72** [,] **3.17** [)] and press ENTER.

```
normalcdf(35,40,
38.72,3.17)
        .5365173032
```

Method 2: To find the probability and include a graph, you must first set up the
WINDOW so that the graph will be displayed properly. You will need to set
Xmin equal to ($\mu - 3\ \sigma$) and Xmax equal to ($\mu + 3\ \sigma$). Press WINDOW and
set **Xmin** equal to ($\mu - 3\ \sigma$) by typing in **38.72 - 3 * 3.17**. Press ENTER and
set **Xmax** equal to ($\mu + 3\ \sigma$) by typing in **38.72 + 3 * 3.17**. Set **Xscl** equal to
σ, which is 3.17.

Setting the Y-range is a little more difficult to do. A good "rule - of - thumb" is to
set **Ymax** equal to .5 / σ. For this example, set **Ymax = .5/3.17**.

Use the blue up arrow to highlight **Ymin**. A good value for **Ymin** is **(-) Ymax / 4**
so type in [(-)] **.158 / 4**.

Press 2nd [DRAW] and select **1:ClrDraw** and press ENTER ENTER. Press 2nd
[STATPLOT] and **TURN OFF** all **PLOTS**. Press 2nd [DISTR], highlight
DRAW and select **1:ShadeNorm(** and type in **35** [,] **40** [,] **38.72** [,] **3.17** [)]
and press ENTER.

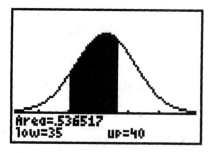

Conclusion: The probability that a randomly selected three-year-old female is between 35 and 40 inches tall is .5365 or 53.65%.

▶ Example 3 (pg. 409) Finding the Value of a Normal Random
Variable

This is an inverse normal problem and the command **invNorm(area,** μ **,** σ **)** is
used. In this type of problem, a percentage of the area under the normal curve is
given and you are asked to find the corresponding X-value. In this example, the
percentage given is the top 80 %. The TI-83 always calculates probability from
negative infinity up to the specified X-value. To find the X-value corresponding to
the top 80%, you must accumulate the bottom 20 % of the area. Press 2ⁿᵈ
[DISTR] and select **3:invNorm(** and type in **.20** [,] **38.72** [,]
3.17) and press ENTER.

```
invNorm(.20,38.7
2,3.17)
        36.05206069
```

Conclusion: The height that separates the top 80% of three-year-old females from
the bottom 20% is 36.05 inches.

◀

▶ Example 4 (pg. 410) Finding the Value of a Normal Random
Variable

This is an inverse normal problem and the command **invNorm(area,** μ **,** σ **)** is
used. In this problem the *middle area* is .98. That leaves an area of .02 to be
equally divided between the *left* and *right* tail areas. Each of these areas are,
therefore, equal to .01. The height that separates the middle 98% from the top 1%
is actually the height that separates the bottom 99% (the middle 98% plus the 1%
in the left tail) from the top 1%.

To find the X-value corresponding to a *left area* of 0.99, press 2nd [DISTR] and
select **3:invNorm(** and type in **.99** , **38.72** , **3.17**) and press ENTER.

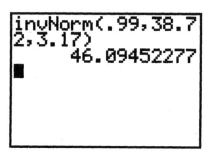

The X-value that separates the top 1% of three-year-old females based on their
heights) from the middle 98% is 46.09 inches.

◀

Section 7.4

▶ Example 1 (pg. 416) A Normal Probability Plot

Press **STAT** and select **1:Edit** and press **ENTER**. Clear all data from **L1.** Enter
the data from Table 4 into **L1**.

To set up the normal probability plot, press **2ⁿᵈ [STAT PLOT]** . Press **ENTER**
to select **Plot 1**. Highlight **On** and press **ENTER**. Set **Type** to the normal
probability plot which is the third selection in the second row. Press **ENTER**.
Set **Data List** to L1 and **Data Axis** to **X**. Next, there are three different types of
Marks that you can select for the graph. The first choice, a small square, is the
best one to use.

Press **ZOOM** and select **9:ZoomStat** and **ENTER**.

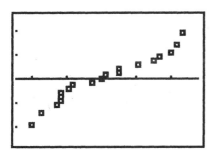

The calculator draws a horizontal line at the X-axis. This plot is *fairly* linear,
indicating that the data generally follows a normal distribution.

Section 7.5

▶ Example 6 (pg. 433) Applying the Central Limit Theorem

The calorie intake of 20-39 year old males is described as a population with a mean, μ, = 2716 and a standard deviation, σ, = 72.8. A sample of 35 males is selected from the population and the calorie intake of each male is recorded. What is the probability that the sample average, $\bar{x}$, is 2750 calories or higher?

Since n > 30, you can conclude that the sampling distribution of the sample mean is approximately normal with $u_{\bar{x}}$ =2716 and $\sigma_{\bar{x}} = 72.8/\sqrt{35}$.

To calculate P($\bar{x} \geq 2750$), press 2nd [DISTR] , select 2:normalcdf(and type in 2750 , 1E99 , 2716 , 72.8/$\sqrt{35}$) and press ENTER.

```
normalcdf(2750,1
E99,2716,72.8/√(
35))
         .0028636574
```

Section 7.6

> ▶ **Example 1 (pg. 440)**   Normal Approximation to the Binomial

In this binomial experiment, the random variable, X, is the number of individuals with blood type O-negative, the probability, p, that an individual has type O-negative blood is .06 and the sample size, n, is 500. We will approximate the probability that fewer than 25 individuals in the sample have type O-negative blood, that is, P(X < 25) using the normal approximation to the binomial.

First, calculate the mean and standard deviation of this binomial random variable. The mean, μ, equals n*p = 500 * .06 = 30. The standard deviation, σ, equals $\sqrt{n * p * (1-p)} = \sqrt{500 * .06 * .94} = 5.31$.

To approximate P(X < 25) with a normal probability we calculate P(X $\leq$ 24.5). (Note: This adjustment from 25 to 24.5 is called a *continuity correction*).

Press 2nd [DISTR] , select **2:normalcdf(** and type in   -1E99 $\boxed{,}$ 24.5 $\boxed{,}$ 30 $\boxed{,}$ 5.31 $\boxed{)}$ and press ENTER.

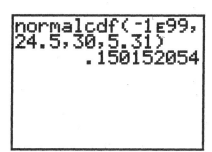

```
normalcdf(-1E99,
24.5,30,5.31)
      .150152054
```

You can compare this probability (.1505) that you obtained through a normal approximation to the actual probability obtained from the binomial distribution.

Press 2nd [DISTR] , select **A:binomcdf(** and type in   500 $\boxed{,}$ .06 $\boxed{,}$ 24 $\boxed{)}$ and press ENTER.

```
binomcdf(500,.06
,24)
       .1493809338
```

The actual $P(X < 25)$ is .1494.

▶ Example 2 (pg. 441) Normal Approximation to the Binomial

In this binomial experiment, the random variable, X, is the number of households with cable TV, the probability, p, that a household has cable TV is .75 and the sample size, n, is 1000. We will approximate the probability that at least **800** households in the sample have cable TV, that is, $P(X \geq 800)$ using the normal approximation to the binomial.

First, calculate the mean and standard deviation of this binomial random variable. The mean, μ, equals n*p = 1000 * .75 = 750. The standard deviation, σ, equals $\sqrt{n * p * (1 - p)} = \sqrt{1000 * .75 * .25} = 13.693$.

To approximate $P(X \geq 800)$ with a normal probability we calculate $P(X \geq 799.5)$ (Note: This adjustment from 800 to 799.5 is called a *continuity correction*).

Press 2ⁿᵈ [DISTR] , select **2:normalcdf(** and type in **799.5** $\boxed{,}$ **1E99** $\boxed{,}$ **750** $\boxed{,}$ **13.693** $\boxed{)}$ and press ENTER.

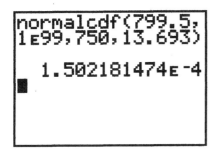

The answer is written in scientific notation. To convert to standard notation, move the decimal point four places to the left. The approximate $P(X \geq 800)$ is .00015, which is an extremely small probability. This suggests that the percentage of households with cable TV is actually higher than 75%.

◀

Confidence Intervals

CHAPTER

8

Section 8.1

▶ Example 3 (pg. 462) Constructing a Z-Interval

Enter the data from Table 1 on pg. 457 into **L1**. Since the sample size is less than 30, we will check for normality using a normal probability plot and we will check for outliers using a Boxplot.

To set up the normal probability plot, press 2nd [STAT PLOT] . Press ENTER to select **Plot 1**. Highlight **On** and press ENTER. Set **Type** to the normal probability plot which is the third selection in the second row. Press ENTER. Set **Data List** to L1 and **Data Axis** to **X**. Next, there are three different types of **Marks** that you can select for the graph. The first choice, a small square, is the best one to use.

Press ZOOM and select **9:ZoomStat** and ENTER.

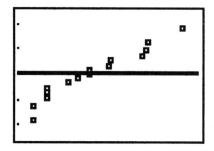

(Note: The calculator draws a horizontal line at the X-axis.) This plot is *fairly* linear, indicating that the data generally follows a normal distribution.

To set up the boxplot, press 2nd [STAT PLOT] . Press ENTER to select **Plot 1**. Highlight **On** and press ENTER. Set **Type** to the boxplot with outliers which is the first selection in the second row. Press ENTER. Set **XList** to **L1** and **Freq** to **1**. Next, there are three different types of **Marks** that you can select for the graph. The first choice, a small square, is the best one to use.

Press ZOOM and select **9:ZoomStat** and ENTER.

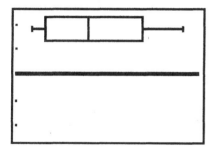

There are no outliers indicated in the boxplot. (Note: Outliers would appear as *'s at the extreme left or right ends of the boxplot.)

Since the data appears to be normally distributed with no outliers, and the population standard deviation is given, the criteria for a Z-interval have been met.

To estimate μ, the population mean, using a 90% confidence interval, press

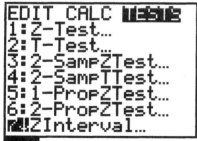

STAT, highlight **TESTS** and select **7:Zinterval.**

On the first line of the display, you can select **Data** or **Stats.** For this example, select **Data** because you want to use the actual data which is in **L1**. Press ENTER . Move to the next line and enter 4100, the assumed value of σ . On the next line, enter **L1** for **LIST**. For **Freq**, enter **1**. For **C-Level**, enter **.90** for a 90% confidence interval. Move the cursor to **Calculate.**

```
ZInterval
 Inpt:DATA Stats
 σ:4100
 List:L₁
 Freq:1
 C-Level:90█
 Calculate
```

Press ENTER .

```
ZInterval
 (36506,39988)
 x̄=38246.93333
 Sx=4521.528036
 n=15
```

A 90% confidence interval estimate of μ, the population mean, is (36506, 39988). The output display includes the sample mean (38246.9), the sample standard deviation (4521.5), and the sample size (15).

> ▶ Problem 13 (pg. 467)

(a.) A random sample of size n = 25 is selected from a population that is normally distributed with a standard deviation, σ, equal to 13. The sample mean, $\bar{x}$, is equal to 108.

To estimate μ, the population mean, using a 96% confidence interval, press

```
EDIT CALC TESTS
1:Z-Test...
2:T-Test...
3:2-SampZTest...
4:2-SampTTest...
5:1-PropZTest...
6:2-PropZTest...
7↓ZInterval...
```

STAT, highlight **TESTS** and select **7:Zinterval.**

On the first line of the display, you can select **Data** or **Stats.** For this example, select **Stats** because you have the sample mean but not the actual data. Press ENTER . Move to the next line and enter **13**, the value of σ . On the next line, enter **108**, the value for $\bar{x}$, the sample mean. On the next line, enter the sample size, **25**. For **C-Level** , enter **.96** for a 96% confidence interval. Move the cursor to **Calculate.**

```
ZInterval
 Inpt:Data Stats
 σ:13
 x̄:108
 n:25
 C-Level:96█
 Calculate
```

Press ENTER .

```
ZInterval
 (102.66,113.34)
 x̄=108
 n=25
```

Section 8.2

▶ Example 2 (pg. 480) Confidence Interval for μ (σ Unknown)

Enter the data from Table 5 into **L1**. Since the sample size is less than 30, the first step is to check for normality using a normal probability plot and then to check for outliers using a Boxplot. (Note: Both steps were done in Example 3, Section 8.1)

In this example, notice that σ is unknown. To construct the confidence interval for μ, the correct procedure under these circumstances (n < 30, σ unknown and the population assumed to be normally distributed) is to use a T-Interval.

Press STAT, highlight **TESTS**, scroll through the options and select **8:TInterval** and press ENTER . Select **Data** for **Inpt** and press ENTER. For **List**, enter **L1** and for **Freq**, enter **1**. Set **C-level** to **.90**. Highlight **Calculate**.

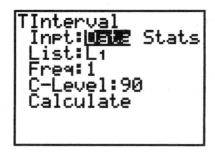

```
TInterval
 Inpt:DATA Stats
 List:L₁
 Freq:1
 C-Level:90
 Calculate
```

Press ENTER .

```
TInterval
 (36191,40303)
 x̄=38246.93333
 Sx=4521.528036
 n=15
```

A 90% confidence interval for μ is (36191, 40303). The sample statistics (mean, standard deviation and sample size are also given in the output screen.

> ▶ Problem 5 (pg. 484)

(a.) In this example, $\bar{x} = 18.4$, s= 4.5 and the sample size, n, = 35.
Since n is greater than 30, and σ, the population standard deviation is unknown, the correct procedure for constructing a confidence interval for μ is the T-procedure.

Press **STAT**, highlight **TESTS**, scroll through the options and select **8:TInterval** and press **ENTER** . In this example, you do not have the actual data. What you have are the summary statistics of the data, so select **Stats** and press **ENTER**. Enter the values for $\bar{x}$, **Sx** and **n.** Enter **.95** for **C-level**. Highlight **Calculate**.

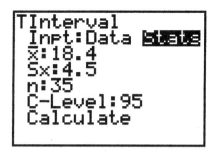

```
TInterval
 Inpt:Data Stats
 x:18.4
 Sx:4.5
 n:35
 C-Level:95
 Calculate
```

Press **ENTER** .

```
TInterval
 (16.854,19.946)
 x=18.4
 Sx=4.5
 n=35
```

A 95% confidence interval estimate for μ is (16.854, 19.946).

▶ Problem 19 (pg. 488)

Enter the data into **L1**. Since the sample size is less than 30, the first step is to check for normality using a normal probability plot and then to check for outliers using a Boxplot.

To set up the normal probability plot, press **2nd [STAT PLOT]** . Press ENTER to select **Plot 1**. Highlight **On** and press ENTER. Set **Type** to the normal probability plot which is the third selection in the second row. Press ENTER. Set **Data List** to **L1** and **Data Axis** to **X**. For **Marks** select the small square.

Press ZOOM and select **9:ZoomStat** and ENTER.

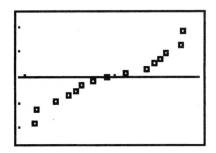

This plot is *fairly* linear, indicating that the data generally follows a normal distribution.

To set up the boxplot, press **2nd [STAT PLOT]** . Press ENTER to select **Plot 1**. Highlight **On** and press ENTER. Set **Type** to the boxplot with outliers which is the first selection in the second row. Press ENTER. Set **XList** to **L1** and **Freq** to **1**. For **Marks** select the small square.

Press ZOOM and select **9:ZoomStat** and ENTER.

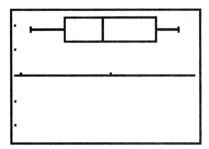

There are no outliers indicated in the boxplot. (Note: Outliers would appear as *'s at the extreme left or right ends of the boxplot.)

Since the data appears to be normally distributed with no outliers, and the population standard deviation is given, the criteria for the Z-interval have been met.

Press **STAT**, highlight **TESTS** and select **7:ZInterval**. Select **Data** for **Inpt** and press **ENTER**. For **List**, enter **L1** and for **Freq**, enter **1**. Set **C-level** to **.95**. Highlight **Calculate.**

```
ZInterval
 Inpt:DATA Stats
 σ:2.9
 List:L₁
 Freq:1
 C-Level:95
 Calculate
```

Press **ENTER**.

```
ZInterval
 (68.386,71.321)
 x̄=69.85333333
 Sx=2.921806356
 n=15
```

▶ Problem 21 (pg. 488)

Enter the data into **L1**. Since the sample size is less than 30, the first step is to check for normality using a normal probability plot and then to check for outliers using a Boxplot.

To set up the normal probability plot, press **2ⁿᵈ** [STAT PLOT] . Press ENTER to select **Plot 1**. Highlight **On** and press ENTER. Set **Type** to the normal probability plot which is the third selection in the second row. Press ENTER. Set **Data List** to **L1** and **Data Axis** to **X**. For **Marks** select the small square.

Press ZOOM and select **9:ZoomStat** and ENTER.

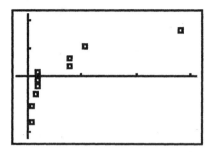

This plot does not look linear. This indicates that the data is NOT normally distributed. Neither the Z-interval nor the T-interval can be used with a small dataset that is not normally distributed.

▶ Problem 23 (pg. 488)

Enter the data into **L1**. Since the sample size is less than 30, the first step is to check for normality using a normal probability plot and then to check for outliers using a Boxplot.

To set up the normal probability plot, press **2ⁿᵈ** [STAT PLOT] . Press ENTER to select **Plot 1**. Highlight **On** and press ENTER. Set **Type** to the normal probability plot which is the third selection in the second row. Press ENTER. Set **Data List** to **L1** and **Data Axis** to **X**. For **Marks** select the small square.

Press ZOOM and select 9:ZoomStat and ENTER.

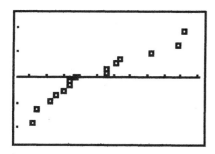

This plot is *fairly* linear, indicating that the data generally follows a normal distribution.

To set up the boxplot, press **2ⁿᵈ** [STAT PLOT] . Press ENTER to select **Plot 1**. Highlight **On** and press ENTER. Set **Type** to the boxplot with outliers which is the first selection in the second row. Press ENTER. Set **XList** to **L1** and **Freq** to **1**. For **Marks** select the small square.

Press ZOOM and select 9:ZoomStat and ENTER.

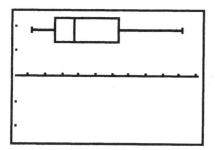

There are no outliers indicated in the boxplot. (Note: Outliers would appear as *'s at the extreme left or right ends of the boxplot.)

Since the data appears to be normally distributed with no outliers, and the population standard deviation is unknown, the criteria for the T-interval have been met.

Press **STAT**, highlight **TESTS** and select **8:T-interval**. Select **Data** for **Inpt** and press **ENTER**. For **List**, enter **L1** and for **Freq**, enter **1**. Set **C-level** to **.95**. Highlight **Calculate.**

```
TInterval
 Inpt:Data Stats
 List:L₁
 Freq:1
 C-Level:95
 Calculate
```

Press **ENTER**.

```
TInterval
 (101.37,117.29)
 x̄=109.3333333
 Sx=14.37590577
 n=15
■
```

Section 8.3

▶ Example 2 (pg. 492) Constructing a Confidence Interval for
a Population Proportion

In this example, 1068 Americans are asked the question: "Have you ever been shot at?" 96 individuals responded 'Yes.' Construct a 95 % confidence interval for p, the true proportion of all Americans who have been shot at.

Press **STAT**, highlight **TESTS**, scroll through the options and select **A:1-PropZInt**. The value for X is the number of American in the group of 1068 who have been shot at, so **X = 96**. The number who were surveyed is n, so **n = 1068**. Enter **.95** for **C-level**.

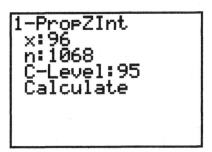

```
1-PropZInt
 x:96
 n:1068
 C-Level:95
 Calculate
```

Highlight **Calculate** and press **ENTER** .

```
1-PropZInt
 (.07273,.10704)
 p̂=.0898876404
 n=1068

■
```

In the output display the confidence interval for p is (.07273, .10704). The sample proportion, $\hat{p}$, is .08989 and the number surveyed is 1068.

Note: You should calculate $n * \hat{p} * (1 - \hat{p})$. This value must be greater than or equal to 10 in order to use this confidence interval procedure. (It is actually easier to do this calculation after you have calculated the confidence interval because the calculator displays the value of $\hat{p}$ as part of the output.). For this example, the calculation is 1068*.09*.91. This value is greater than 10, so this supports the use of the confidence interval procedure.

Hypothesis Testing

Section 9.2

▶ Example 3 (pg. 535) Computing the P-value of a Right-Tailed Test

Enter the data from Table 1 on pg. 532 into L1. Because the sample size is less than 30, the data must be tested for normality and checked for outliers.

To set up the normal probability plot, press **2ⁿᵈ [STAT PLOT]**. Press ENTER to select **Plot 1**. Highlight **On** and press ENTER. Set **Type** to the normal probability plot which is the third selection in the second row. Press ENTER. Set **Data List** to **L1** and **Data Axis** to **X**. For **Marks** select the small square.

Press ZOOM and select **9:ZoomStat** and ENTER.

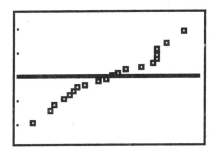

This plot is *fairly* linear, indicating that the data generally follows a normal distribution.

To set up the boxplot, press **2ⁿᵈ [STAT PLOT]**. Press ENTER to select **Plot 1**. Highlight **On** and press ENTER. Set **Type** to the boxplot with outliers which is the first selection in the second row. Press ENTER. Set **XList** to **L1** and **Freq** to **1**. For **Marks** select the small square.

Press ZOOM and select **9:ZoomStat** and ENTER.

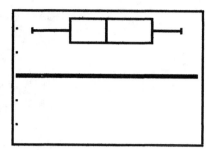

There are no outliers indicated in the boxplot. (Note: Outliers would appear as *'s at the extreme left or right ends of the boxplot.)

The hypothesis test, $H_o : \mu = 10300$ vs. $H_a : \mu > 10300$, is a right-tailed test. The population standard deviation, σ, is 3500. The **Z-Test** is the appropriate test. To run the test, press **STAT**, highlight **TESTS** and select **1:Z-Test**. Since you are using the actual data, which is stored in L1, for the analysis, select **Data** for **Inpt** and press **ENTER**. For μ_0 enter 10300, the value for μ in the null hypothesis. For σ enter 3500. Enter **L1** for **List,** and **1** for **Freq**. On the next line, choose the appropriate alternative hypothesis and press **ENTER**. For this example, it is $> \mu_0$, a right-tailed test.

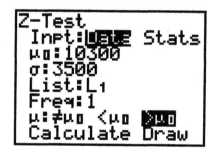

There are two choices for the output of this test. The first choice is **Calculate**. The output displays the alternative hypothesis, the calculated z-value, the P-value, $\bar{x}$ and n.

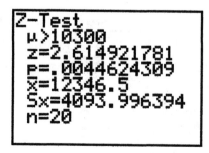

Since p = .0045, which is less than α, the correct conclusion is to **Reject** H_o.

To view the second output option, press STAT, highlight **TESTS**, and select **1:Z-Test**. All the necessary information for this example is still stored in the calculator. Scroll down to the bottom line and select **DRAW**. A normal curve is displayed with the right-tail area of .0045 shaded. (Note: Because the area is so small in this example, it is not really visible in the curve.) This shaded area is the area to the right of the calculated Z-value. The Z-value and the P-value are also displayed.

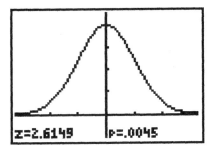

▶ Example 4 (pg. 536)   Computing the P-value of a Two-Tailed Test

Enter the data from Table 2 on pg. 533 into L1. Because the sample size is less than 30, the data must be tested for normality and checked for outliers.

To set up the normal probability plot, press **2ⁿᵈ [STAT PLOT]** . Press ENTER to select **Plot 1**. Highlight **On** and press ENTER. Set **Type** to the normal probability plot which is the third selection in the second row. Press ENTER. Set **Data List** to **L1** and **Data Axis** to **X**. For **Marks** select the small square.

Press ZOOM and select **9:ZoomStat** and ENTER.

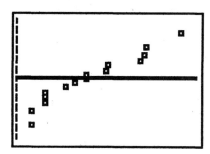

This plot is *fairly* linear, indicating that the data generally follows a normal distribution.

To set up the boxplot, press **2ⁿᵈ [STAT PLOT]** . Press ENTER to select **Plot 1**. Highlight **On** and press ENTER. Set **Type** to the boxplot with outliers which is the first selection in the second row. Press ENTER. Set **XList** to **L1** and **Freq** to **1**. For **Marks** select the small square.

Press ZOOM and select **9:ZoomStat** and ENTER.

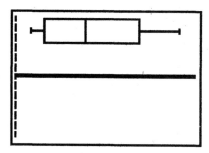

There are no outliers indicated in the boxplot. (Note: Outliers would appear as *'s at the extreme left or right ends of the boxplot.)

This test is a two-tailed test for $H_o : \mu = 37500$ vs. $H_a : \mu \neq 37500$. The population standard deviation, σ, is 4100. The **Z-Test** is the appropriate test. To run the test, press STAT, highlight **TESTS** and select **1:Z-Test**. Since you are using the actual data, which is stored in L1, for the analysis, select **Data** for **Inpt** and press ENTER. For μ_0 enter 37500, the value for μ in the null hypothesis. For σ enter **4100**. Enter **L1** for **List,** and **1** for **Freq**. On the next line, choose the appropriate alternative hypothesis and press ENTER. For this example, it is $\neq \mu_0$, a two-tailed test.

```
Z-Test
 Inpt:DATA Stats
 µ0:37500
 σ:4100
 List:L1
 Freq:1
 µ:≠µ0 <µ0 >µ0
 Calculate Draw
```

Highlight **Calculate** and press ENTER .

```
Z-Test
 µ≠37500
 z=.7055756977
 P=.4804518931
 x̄=38246.93333
 Sx=4521.528036
 n=15
 ▮
```

Or, highlight **Draw** and press ENTER.

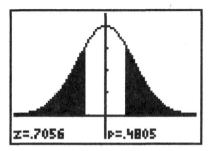

```
z=.7056        P=.4805
```

Notice the P-value is equal to .4805. In this example, α is .01. Since the P-value is greater than α, the correct conclusion is to **Fail to Reject** H_o. (Note: The P-value calculated using the TI-83 is slightly different from the P-value obtained using the Z-table. That difference is simply due to rounding.)

▶ Example 5 (pg. 538) Using a Confidence Interval to Test a
Hypothesis

Enter the data from Table 2 on pg. 533 into L1. Because the sample size is less
than 30, the data must be tested for normality and checked for outliers. (Note:
These tests were done with the previous Example and the results indicated that the
data was normally distributed with no outliers.)

To estimate μ, the population mean, using a 90% confidence interval, press
STAT, highlight **TESTS** and select **7:Zinterval.**

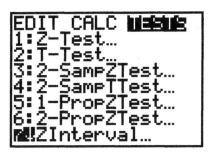

On the first line of the display select **Data**. Press **ENTER** . Move to the next line
and enter 4100, the assumed value of σ. On the next line, enter **L1** for **LIST**.
For **Freq**, enter **1**. For **C-Level**, enter **.90** for a 90% confidence interval. Move
the cursor to **Calculate.**

```
ZInterval
 (36506,39988)
 x̄=38246.93333
 Sx=4521.528036
 n=15
```

The 90% confidence interval for μ is (36506, 39988). Notice that this confidence
interval contains the hypothesized value for μ (37500). Since the hypothesized
value is contained in the confidence interval, the correct decision is: **Fail to Reject
the null hypothesis.**

◀

▶ Problem 5 (pg. 539)

Test the hypotheses: $H_o : \mu = 20$ vs. $H_a : \mu < 20$. The underlying population is assumed to be normally distributed with $\sigma = 3$. The sample mean, $\bar{x}$, = 18.3, and n = 18. Press **STAT**, highlight **TESTS** and select **1:Z-Test**. For **Inpt**, choose **Stats** and press **ENTER**. Fill the input screen with the appropriate information. Choose < μ_0 for the alternative hypothesis and press **ENTER**.

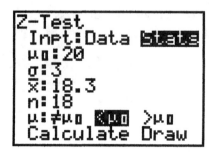

Highlight **Calculate** and press **ENTER**.

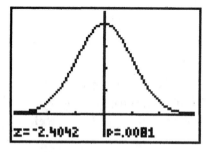

Or, highlight **Draw** and press **ENTER**.

The P=value is .008. So, less than 1 sample in 100 will result in a sample mean of 18.3 or less, if, in fact, the population mean is equal to 20. Since the P-value is less than α, the correct conclusion is to **Reject** H_o.

▶ Problem 7 (pg. 539)

Test the hypotheses: $H_o : \mu = 105$ vs. $H_a : \mu \neq 105$. In this example, the sample size is greater than 30. Since the sample size is *large*, the Central Limit Theorem applies and we can assume that the sampling distribution of $\bar{x}$ is approximately normal. The population standard deviation, σ, is equal to 12.

To run the test, press **STAT**, highlight **TESTS** and select **1:Z-Test**. For **Inpt**, choose **Stats** and press **ENTER**. Fill the input screen with the appropriate information. Choose $\neq \mu_0$ for the alternative hypothesis and press **ENTER**.

Highlight **Calculate** and press **ENTER**.

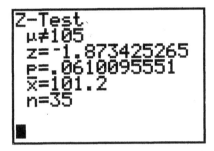

Or, highlight **Draw** and press **ENTER**.

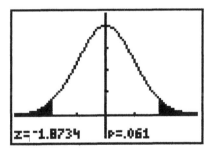

The P=value is .061. Since the P-value is greater than α, the correct conclusion is to **Fail to Reject** H_o.

Section 9.3

▶ Example 2 (pg. 549)   Calculating P-values for a T-Test

Enter the data from Table 3 on pg. 548 into L1. Because the sample size is less than 30, the data must be tested for normality and checked for outliers.

To set up the normal probability plot, press **2ⁿᵈ [STAT PLOT]** . Press ENTER to select **Plot 1**. Highlight **On** and press ENTER. Set **Type** to the normal probability plot which is the third selection in the second row. Press ENTER. Set **Data List** to **L1** and **Data Axis** to **X**. For **Marks** select the small square.

Press ZOOM and select **9:ZoomStat** and ENTER.

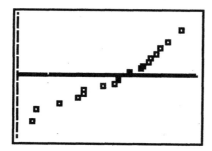

This plot is *fairly* linear, indicating that the data generally follows a normal distribution.

To set up the boxplot, press **2ⁿᵈ [STAT PLOT]** . Press ENTER to select **Plot 1**. Highlight **On** and press ENTER. Set **Type** to the boxplot with outliers which is the first selection in the second row. Press ENTER. Set **XList** to **L1** and **Freq** to 1. For **Marks** select the small square.

Press ZOOM and select **9:ZoomStat** and ENTER.

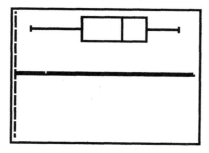

There are no outliers indicated in the boxplot. (Note: Outliers would appear as *'s at the extreme left or right ends of the boxplot.)

This test is a right-tailed test of $H_o : \mu = 142.8$ vs. $H_a : \mu > 142.8$. Since $n < 30$, and the population standard deviation, σ, is unknown, the T-Test is the appropriate test. This test requires the underlying population to be approximately normally distributed with no outliers, as was verified in the plots.

Press **STAT**, highlight **TESTS** and select **2:T-Test**. Choose **Data** for **Inpt** and press **ENTER**. Fill in the following information: μ_0 = **142.8, List = L1,** and **Freq =1**. Choose the right-tailed alternative hypothesis, $> \mu_0$, and press **ENTER**. Highlight **Calculate** and press **ENTER**

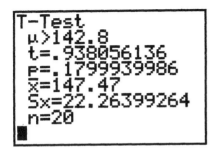

Or, highlight **Draw** and press **ENTER**.

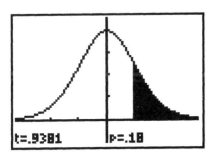

Since the P-value is greater than α, the correct conclusion is to **Fail to Reject** H_o.

> ▶ Problem 5 (pg. 551)

Test the hypotheses: $H_o: \mu = 20$ vs. $H_a: \mu < 20$.   The underlying population is known to be normally distributed. The sample statistics are $\bar{x} = 18.3$, s = 4.3 and n = 18. Press **STAT**, highlight **TESTS** and select **2:T-Test**. For **Inpt**, choose **Stats** and press **ENTER**. Fill the input screen with the appropriate information. Choose $< \mu_0$ for the alternative hypothesis and press **ENTER**.

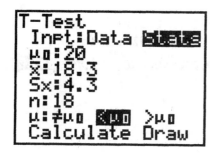

Highlight **Calculate** and press **ENTER**.

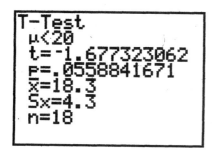

Or, highlight **Draw** and press **ENTER**.

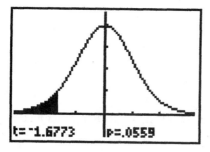

The P=value is .0559. So, approximately 5 samples in 100 will result in a sample mean of 18.3 or less, if, in fact, the population mean is equal to 20. Since the P-value is greater than α, the correct conclusion is to **Fail to Reject** H_o.

Section 9.4

▶ Example 2 (pg. 562)   Side Effects of Prevnar - Hypothesis Test
                        For a Proportion

This hypothesis test is a two-tailed test of: $H_o : p = .135$ vs. $H_a : p \neq .135$. The procedure that is used for this test is the **1-Proportion Test**. This test has two requirements. The first requirement is: $n * p_0 * (1 - p_0) \geq 10$. To verify this, calculate 710*.135*(1-.135). The result is greater than 10, so the first requirement is satisfied. The second requirement is that the sample size is not more than 5% of the *population* size. In this example, the population is *all babies between 12 and 15 months of age*. We don't know the exact size of the population, but it is in the millions. The sample size of 710 is definitely less than 5% of the population size.

To run the test, press **STAT**, highlight **TESTS** and select **5:1-PropZTest**. This test requires a value for p_0, which is the value for p in the null hypothesis. Enter **.135** for p_0. Next, a value for X is required. X is the number of "successes" in the sample. In this example, a success is " experiencing a decrease in appetite", so **X** is equal to **121**. Next, enter the value for **n**. Select $\neq p_0$ for the alternative hypothesis and press **ENTER**.

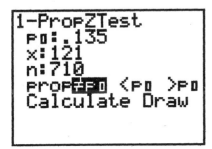

```
1-PropZTest
 p0:.135
 x:121
 n:710
 prop≠p0 <p0 >p0
 Calculate Draw
```

Highlight **Calculate** and press **ENTER**.

```
1-PropZTest
 prop≠.135
 z=2.762064888
 p=.0057438195
 p̂=.1704225352
 n=710
```

The output displays the alternate hypothesis that was selected, the calculated Z-value, the P-value, the sample proportion, $\hat{p}$, and the sample size. (Note: $\hat{p} = 121/710$.)

Or, highlight **Draw** and press ENTER.

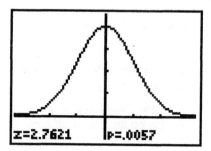

Since the P-value is less than α, the correct conclusion is to **Reject** H_o.

> ▸ Example 3 (pg. 564) Hypothesis Test for a Proportion – Small Sample Case

In this test of a population proportion, the requirement $n * p_0 * (1 - p_0) \geq 10$ is not satisfied. (The calculation $35*.489*(1-.489)$ is equal to 8.75.) (Note: In cases in which the sample size is relatively small this requirement is often not satisfied.)

An alternative method of testing a hypothesis about a population proportion is to use the binomial probability formula to calculate the likelihood of the sample result. If the sample result is *unusual* then we will **reject the null hypothesis**. We define *unusual* events as events that have a probability less than .05.

The hypothesis test is: $H_o : p = .489$ vs. $H_a : p > .489$. The sample statistics are $n = 35$ and $X = 21$. Using the binomial probability formula, we calculate the likelihood of obtaining 21 or more males who consume the recommended daily allowance of calcium in the sample of 35. We assume that the proportion of males in the population who consume the recommended daily allowance of calcium is .489.

Press $1 - 2^{nd}$ DISTR and select **A:binomcdf(.** Type in **35 , .489 , 20).**

```
1-binomcdf(35,.4
89,20)
      .1261068497
```

(Note: The command **binomcdf** calculates the probability that $X \leq 20$, which is the *complement* of the probability that $X \geq 21$. To obtain $P(X \geq 21)$, we calculate $P(X \leq 20)$ and subtract this value from 1.)

The result is: $P(X \geq 21) = .126$. Since this probability is greater than .05, the correct conclusion is to **Fail to Reject** H_o.

◀

> ▶ Problem 21 (pg. 566)

This hypothesis test is a left-tailed test of: $H_o: p = .37$ vs. $H_a: p < .37$. To use the **1-Proportion Test,** first you must determine whether the requirements for this test have been satisfied. The first requirement is: $n * p_0 * (1 - p_0) \geq 10$. To verify this, calculate 150*.37*(1-.37). The result is greater than 10, so the first requirement is satisfied. The second requirement is that the sample size is not more than 5% of the *population* size. In this example, the population is *all pet owners.* We don't know the exact size of the population, but it is in the millions. The sample size of 150 is definitely less than 5% of the population size.

To run the test, press **STAT**, highlight **TESTS** and select **5:1-PropZTest**. Enter .37 for p_0. For **X,** enter **54** and for n, enter **150**. Select $< p_0$ for the alternative hypothesis and press **ENTER**.

Highlight **Calculate** and press **ENTER**.

```
1-PropZTest
 prop<.37
 z=-.2536731447
 P=.3998740943
 p=.36
 n=150
```

Or highlight **Draw** and press **ENTER**.

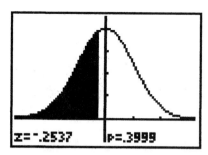

```
z=-.2537        p=.3999
```

Since the P-value is greater than α, the correct conclusion is to **Fail to Reject** H_o. The veterinarian's claim that less than 37% of pet owners speak to their pets is not supported by the data.

◀

▶ Problem 23 (pg. 567)

In this test of a population proportion, the requirement $n * p_0 * (1 - p_0) \geq 10$ is not satisfied. (The calculation $120*.04*(1-.04)$ is equal to 4.608.) (Note: In cases in which the sample size is relatively small this requirement is often not satisfied.)

An alternative method of testing a hypothesis about a population proportion is to use the binomial probability formula to calculate the likelihood of the sample result. If the sample result is *unusual* then we will **reject the null hypothesis**. We define *unusual* events as events that have a probability less than .05.

The hypothesis test is: $H_o : p = .04$ vs. $H_a : p < .04$. The sample statistics are $n = 120$ and $X = 3$. Using the binomial probability formula, we calculate the likelihood of obtaining 3 or fewer mothers who smoked 21 or more cigarettes during pregnancy. We assume that the proportion of mothers who smoked 21 or more cigarettes during pregnancy is .04.

Press 2nd DISTR and select **A:binomcdf(**. Type in **120 , .04 , 3)**.

```
binomcdf(120,.04
,3)
        .288658855
```

The result is: $P(X \leq 3) = .2887$. Since this probability is greater than .05, the correct conclusion is to **Fail to Reject** H_o. The data does not support the obstetrician's belief that less than 4% of mothers smoked 21 or more cigarettes during pregnancy.

◀

Section 9.5

▶ Example 1 (pg. 570) Testing a Hypothesis About a Standard Deviation

This is a right-tailed hypothesis test about σ. The hypotheses are: H_o: σ = 3500 vs. H_a: σ > 3500.

Enter the data from Table 4 on pg. 73 into L1. Because the sample size is less than 30, the data must be tested for normality and checked for outliers.

To set up the normal probability plot, press 2ⁿᵈ [STAT PLOT] . Press ENTER to select **Plot 1**. Highlight **On** and press ENTER. Set **Type** to the normal probability plot which is the third selection in the second row. Press ENTER. Set **Data List** to **L1** and **Data Axis** to **X**. For **Marks** select the small square.

Press ZOOM and select 9:ZoomStat and ENTER.

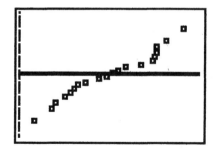

This plot is *fairly* linear, indicating that the data generally follows a normal distribution.

To set up the boxplot, press 2ⁿᵈ [STAT PLOT] . Press ENTER to select **Plot 1**. Highlight **On** and press ENTER. Set **Type** to the boxplot with outliers which is the first selection in the second row. Press ENTER. Set **XList** to **L1** and **Freq** to **1**. For **Marks** select the small square.

Press ZOOM and select 9:ZoomStat and ENTER.

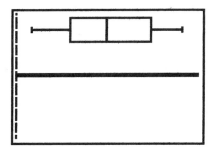

There are no outliers indicated in the boxplot. (Note: Outliers would appear as *'s at the extreme left or right ends of the boxplot.)

The next step is to find the value of the sample standard deviation. Press **STAT**, highlight **CALC** and select **1:1-Var Stats** and press **ENTER**. Type in **L1** and press **ENTER**. The sample statistics are displayed on the screen. The sample standard deviation, **Sx = 4094**.

Next, calculate the test statistic: $\chi^2 = (n-1)s^2 / \sigma_0^2$. The calculation, $19(4094)^2/3500^2 = 25.996$. Because this is a right-tailed test, to find the P-value associated with the test statistic value of 25.996, we calculate the area under the χ^2-curve to the *right* of 25.996. Press **2nd DISTR**, and select **7: χ^2 cdf(**. This calculation requires a *lowerbound*, an *upperbound* and the *degrees of freedom*. The lowerbound is 25.996, the upperbound is positive infinity (1E99 on the calculator) and the degrees of freedom value is 19 (which is equal to n-1).

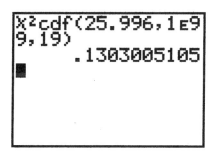

Since the P-value of .1303 is greater than α, the correct conclusion is to **Fail to Reject** H_o. There is not sufficient evidence to support that claim that $\sigma > 3500$.

▶ Problem 5 (pg. 573)

This is a two-tailed hypothesis test about σ. The hypotheses are: H_o: σ = 4.3 vs. H_a: σ ≠ 4.3. The population is normally distributed, the sample size is 12 and the sample standard deviation is 4.8.

The first step is to calculate the test statistic: $\chi^2 = (n-1)s^2 / \sigma_0^2$. The calculation, $11(4.8)^2/4.3^2 = 13.71$. To find the P-value associated with the test statistic value of 13.71, we calculate the area under the χ^2-curve to the *right* of 13.71. The reason that we are finding the area *to the right* of 13.71 is because the sample standard deviation, 4.8, is *greater than* 4.3. Press 2nd DISTR, and select 7: χ^2 **cdf(**. This calculation requires a *lowerbound*, an *upperbound* and the *degrees of freedom*. The lowerbound is 13.71, the upperbound is positive infinity (1E99 on the calculator) and the degrees of freedom value is 11 (which is equal to n-1).

```
X²cdf(13.71,1E99
,11)
        .2494577162
```

The p-value is .2495. Because this is a two-tailed test, you must compare this P-value to α/2 which is .025. Since the P-value is greater than .025, the correct decision is to **Fail to reject** H_o. There is not sufficient evidence to support that claim that σ ≠ 4.3.

◀

Inferences on Two Samples

Section 10.1

▶ Example 2 (pg. 596) Testing a Claim Regarding Matched Pairs Data

In this example, the data (found in Table 1 on pg. 597) is paired data, with two reaction times for each of the students. Enter the reaction times for the individual's dominant hand in L1 and enter the reaction times for the individual's non-dominant hand into L2. Next, you must create a set of differences, d = reaction time of dominant hand - reaction time of non-dominant hand. To create this set, move the cursor to highlight the label **L3,** found at the top of the third column, and press ENTER. Notice that the cursor is flashing on the bottom line of the display. Press 2nd [L1] - 2nd [L2]

L1	L2	L3	3
.177	.179	------	
.21	.202		
.186	.208		
.189	.184		
.198	.215		
.194	.193		
.16	.194		

L3 =L₁−L₂

and press ENTER.

L1	L2	L3	3
.177	.179	-.002	
.21	.202	.008	
.186	.208	-.022	
.189	.184	.005	
.198	.215	-.017	
.194	.193	.001	
.16	.194	-.034	

L3(1)= -.002

Each value in **L3** is the difference **L1 - L2**.

To test the claim that the reaction time in an individual's dominant hand is less than the reaction time in his/her non-dominant hand, the hypothesis test is:
$H_o: \mu_d = 0$ vs. $H_a: \mu_d < 0$.

Because the sample size is less than 30, the set of differences must be tested for normality and checked for outliers.

To set up the normal probability plot, press 2nd [STAT PLOT] . Press ENTER to select **Plot 1**. Highlight **On** and press ENTER. Set **Type** to the normal probability plot which is the third selection in the second row. Press ENTER. Set **Data List** to **L3** and **Data Axis** to **X**. For **Marks,** select the small square.

Press ZOOM and select **9:ZoomStat** and ENTER.

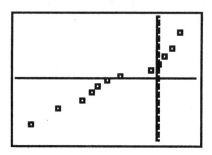

This plot is *fairly* linear, indicating that the data generally follows a normal distribution.

To set up the boxplot, press 2nd [STAT PLOT] . Press ENTER to select **Plot 1**. Highlight **On** and press ENTER. Set **Type** to the boxplot with outliers which is the first selection in the second row. Press ENTER. Set **XList** to **L3** and **Freq** to 1. For **Marks** select the small square.

Press ZOOM and select **9:ZoomStat** and ENTER.

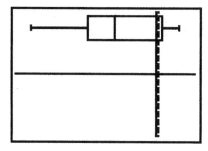

Note: Both of these graphs include the X-axis and Y-axis with the plot. To turn the axes off, press GRAPH . Next press 2nd [FORMAT] and scroll down to

Axes On. Move the cursor to **Axes Off** and press ENTER . If you redo the graphs, the axes will no longer appear on the screen. You may prefer the way the graphs look without the axes. Here is the graph of the Boxplot:

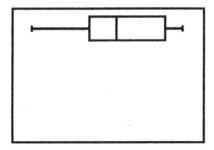

There are no outliers indicated in the boxplot. (Note: Outliers would appear as *'s at the extreme left or right ends of the boxplot.)

To run the hypothesis test, press STAT, highlight **TESTS** and select **2:T-Test**. In this example, you are using the actual data to do the analysis, so select **Data** for **Inpt** and press ENTER. The value for μ_o is **0**, the value in the null hypothesis. The set of differences is found in **L3**, so set **List** to **L3**. Set **Freq** equal to **1**. Choose $< \mu_o$ as the alternative hypothesis and highlight **Calculate** and press ENTER.

```
T-Test
 μ<0
 t=-2.775932838
 P=.0090173095
 x=-.0131666667
 Sx=.0164307546
 n=12
```

You can also highlight **DRAW** and press ENTER.

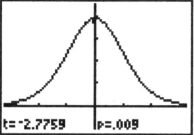

Since the P-value is less than α, the correct decision is to **Reject** H_o. We conclude that an individual's dominant hand has a faster reaction time.

▶ Example 3 (pg. 600) Constructing a Confidence Interval for
Matched-Pairs Data

In Example 2 we created a column of differences in L3. We also confirmed that the data
was normally distributed with no outliers. To create a 95% confidence interval for the
mean difference, press STAT, highlight **TESTS** and select **8:T-Interval**. In this
example, you are using the actual data to do the analysis, so select **Data** for **Inpt**
and press ENTER. The set of differences is found in **L3**, so set **List** to **L3**. Set
Freq equal to **1**. Set **C-Level** to **.95**. Highlight **Calculate** and press ENTER.

```
TInterval
 (-.0236,-.0027)
 x̄=-.0131666667
 Sx=.0164307546
 n=12
```

The 95% confidence interval for μ_d, the mean difference in reaction time, is
(-.0236, -.0027) seconds. We interpret this interval to mean that, on average, the
reaction time of a person's dominant hand is between .0027 and .0236 seconds
less than the reaction time of a person's non-dominant hand.

◀

▶ Problem 13 (pg. 603)

In this example the data is paired data, with two measurements of water clarity at the same location in the lake at Joliet Junior College. The first reading is taken at a specific location on the lake at a particular time during a given year. The second reading is taken at the same location 5 years later. Enter the **initial depth** readings into L1 and the readings **5 years later** into L2. Next, you must create a set of differences, d = intial depth reading - reading 5 years later. To create this set, move the cursor to highlight the label **L3,** found at the top of the third column, and press ENTER. Notice that the cursor is flashing on the bottom line of the display. Press 2nd [L1] - 2nd [L2] and press ENTER.

L1	L2	L3 3
38	52	-14
58	60	-2
65	72	-7
74	72	2
56	54	2
36	48	-12
56	58	-2
L3(1)= -14		

Each value in L3 is the difference L1 - L2.

We are interesting in testing the claim that the clarity of the lake is improving. If the clarity of the lake *is improving*, then the depth at which the disk is no longer visible should be getting deeper. If that is the case, then the difference, **L1 – L2**, should be *negative*. In other words, the original depths that were measured should be "less deep" than the measurements 5 years later. So the appropriate alternate hypothesis is: $\mu_d < 0$.

Note: A normal probability plot and boxplot of the data indicate that the differences are approximately normal with no outliers.

(a.) To run the hypothesis test, press STAT, highlight **TESTS** and select **2:T-Test.** In this example, you are using the actual data to do the analysis, so select **Data** for **Inpt** and press ENTER. The value for μ_o is **0**, the value in the null hypothesis. The set of differences is found in **L3,** so set **List** to **L3.** Set **Freq** equal to **1.** Choose $< \mu_o$ as the alternative hypothesis and highlight **Calculate** and press ENTER.

```
T-Test
 μ<0
 t=-2.383651869
 P=.0243101879
 x̄=-5.125
 Sx=6.081294505
 n=8
```

Or, highlight **Draw** and press ENTER.

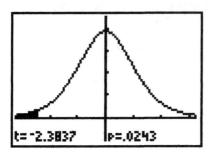

Since the P-value is less than α, the correct decision is to **Reject** H_o. We conclude that the clarity of the lake is improving.

(b.) Press STAT, highlight TESTS and select **8:T-Interval**. Select **Data** for **Inpt,** set **List** to **L3,** set **Freq** equal to **1**and set **C-Level** to **.95**. Highlight **Calculate** and press ENTER.

```
TInterval
 (-10.21, -.0409)
 x̄=-5.125
 Sx=6.081294505
 n=8

■
```

The average difference in the depth at which the disk is visible is (-.04, -10.21) feet. This means that the average depth at which the disk is visible has increased by an amount between .04 and 10.21 feet.

◀

▶ Problem 17 (pg. 605)

In this example the data is paired data consisting of daily car rental fees for two different companies at 10 locations in the United States. Enter the rental rates for *Thrifty* into **L1** and the rental rates for *Hertz* into **L2**. Next, create a set of differences, d = *Thrifty - Hertz*. Move the cursor to highlight the label **L3**, found at the top of the third column, and press **ENTER**. Press 2nd **[L1]** - 2nd **[L2]** and press **ENTER**. Each value in **L3** is the difference **L1 - L2**.

We are interested in testing the claim that *Thrifty* is less expensive than *Hertz*. If that is the case, then the difference, **L1 − L2**, should be *negative*. So the appropriate alternate hypothesis is: $\mu_d < 0$.

Note: A normal probability plot and boxplot of the data indicate that the differences are approximately normal with no outliers.

(a.) To run the hypothesis test, press **STAT**, highlight **TESTS** and select **2:T-Test**. In this example, you are using the actual data to do the analysis, so select **Data** for **Inpt** and press **ENTER**. The value for μ_o is **0**, the value in the null hypothesis. The set of differences is found in **L3**, so set **List** to **L3**. Set **Freq** equal to **1**. Choose $< \mu_o$ as the alternative hypothesis and highlight **Calculate** and press **ENTER**.

```
T-Test
 μ<0
 t=.0890092826
 p=.5344881153
 x̄=.259
 Sx=9.201623589
 n=10
■
```

Or, highlight **Draw** and press **ENTER**.

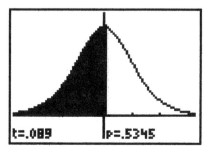

Since the P-value is greater than α , the correct decision is to **Fail to Reject** H_o.
We cannot conclude that *Thrifty* is less expensive than *Hertz*.

(b.) Press **STAT**, highlight **TESTS** and select **8:T-Interval**. Select **Data** for
Inpt, set **List** to **L3,** set **Freq** equal to 1and set **C-Level** to **.95**. Highlight
Calculate and press **ENTER**.

```
TInterval
 (-5.075,5.593)
 x̄=.259
 Sx=9.201623589
 n=10
```

The average difference in price is between (–5.08 and 5.59) dollars. Since this
confidence interval contains 0, we conclude that there is no difference in average price.

Section 10.2

▶ Example 1 (pg. 610) Testing a Claim Regarding Two Means

To test the claim that the flight animals have a different red blood cell mass than the control animals, use a two-tailed test: $H_o: \mu_1 = \mu_2$ vs $H_a: \mu_1 \neq \mu_2$. Enter the data from the 14 flight rats into **L1** and the data from the control rats into **L2**. Because the data sets are small, the first step is to use normal probability plots and boxplots to verify that both datasets are approximately normal and contain no outliers.

To run the hypothesis test, press **STAT**, highlight **TESTS**, and select **4:2-SampTTest**. Since you are inputting the sample data, select **Data** and press **ENTER**. Enter **L1** for **List1** and **L2** for **List 2**. Set **Freq1** and **Freq2** to 1. Select $\neq \mu_2$ as the alternative hypothesis and press **ENTER**. Scroll down to the next line. On this line, there are two options. Select **NO** because, in the procedure we are using (called Welch's approximate t-test), the variances are NOT assumed to be equal and therefore, we do not want a pooled variance. Press **ENTER**.

```
2-SampTTest
 Inpt:Data Stats
 List1:L₁
 List2:L₂
 Freq1:1
 Freq2:1
 µ1:≠µ2 <µ2 >µ2
↓Pooled:No Yes
```

Scroll down to the next line, highlight **Calculate** and press **ENTER**.

```
2-SampTTest
 µ1≠µ2
 t=-1.436781704
 P=.1627070766
 df=25.99635232
 x̄1=7.880714286
↓x̄2=8.43
```

The output (shown above) for **Calculate** displays the alternative hypothesis, the test statistic, the P-value, the degrees of freedom and the sample statistics.

Notice the degrees of freedom = 25.996. In cases, such as this one, in which the population variances are not assumed to be equal, the calculator calculates an adjusted degrees of freedom, (see the formula on pg. 612 of your textbook.)

If you choose **Draw**, the output includes a graph with the area associated with the P-value shaded.

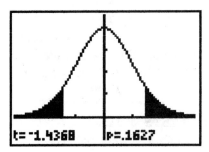

Since the P-value is greater than α, we fail to reject H_0. There is not sufficient evidence to support the claim that there is a significant difference in the red blood cell mass of the flight animals and the control animals.

▶ Example 2 (pg. 613) Constructing a Confidence Interval about the Difference of Two Means

For this example, we will use the information from Table 4 on pg. 612. In Example 1, we tested for normality using normal probability plots and for outliers using Boxplots. Both data sets appeared to be approximately normal with no outliers.

To construct a 95% confidence interval for $(\mu_1 - \mu_2)$, press **STAT**, highlight **TESTS** and select **0:2-SampTint**. For **Inpt**, select **Stats** and press **ENTER**. Enter the mean of the first sample, 7.881, the standard deviation, 1.017, and the sample size, 14. Then enter the mean (8.43), standard deviation (1.005), and sample size (14) of the second sample. Scroll down to the next line and enter .95 for the **C-level**. On the next line, select **No**, because we are not using a pooled variance.

```
2-SampTInt
 Inpt:Data Stats
 x̄1:7.881
 Sx1:1.017
 n1:14
 x̄2:8.43
 Sx2:1.005
↓n2:14
```

Scroll down to the next line, **Calculate,** and press **ENTER**

```
2-SampTInt
 (-1.334,.23648)
 df=25.9963378
 x̄₁=7.881
 x̄₂=8.43
 Sx₁=1.017
↓Sx₂=1.005
```

A 95 % confidence interval for the difference in the population means is (-1.334, .23648). Since this interval contains 0, the correct conclusion is that there is no difference in the red blood cell mass of the two groups. (Note: This interval differs slightly from the textbook interval because the calculator uses the adjusted degrees of freedom in the calculations.)

◀

▶Problem 7 (pg. 615)

To test the claim that the treatment group experienced a larger mean improvement than the control group, use a one-tailed test: $H_o: \mu_1 = \mu_2$ vs H_a: $\mu_1 > \mu_2$. Because the data sets are large (n_1 and $n_2 > 30$), we know that the sampling distribution of ($\bar{x}_1 - \bar{x}_2$) is approximately normal and we can safely use the Two-sample T-test.

To run the hypothesis test, press **STAT**, highlight **TESTS**, and select **4:2-SampTTest**. Since you are inputting the sample statistics, select **Stats** and press **ENTER**. Enter the sample statistics for the two samples.

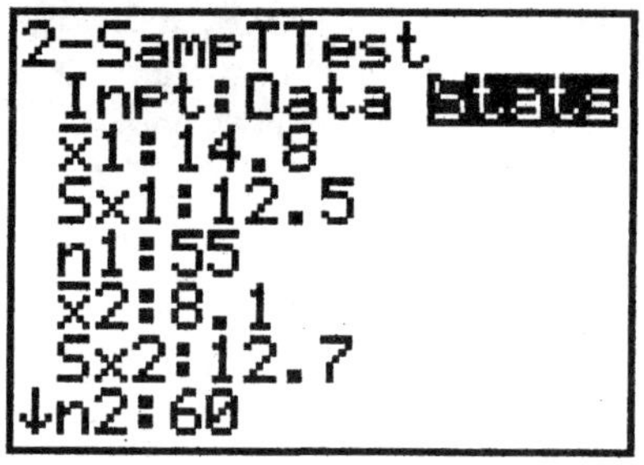

```
2-SampTTest
 Inpt:Data Stats
 x1:14.8
 Sx1:12.5
 n1:55
 x2:8.1
 Sx2:12.7
↓n2:60
```

Scroll down to the next line, select $>\mu_2$ as the alternative hypothesis and press **ENTER**. Scroll down to the next line. Select **NO** because we are not using a pooled variance. Press **ENTER**.

Scroll down to the next line, highlight **Calculate** and press **ENTER**.

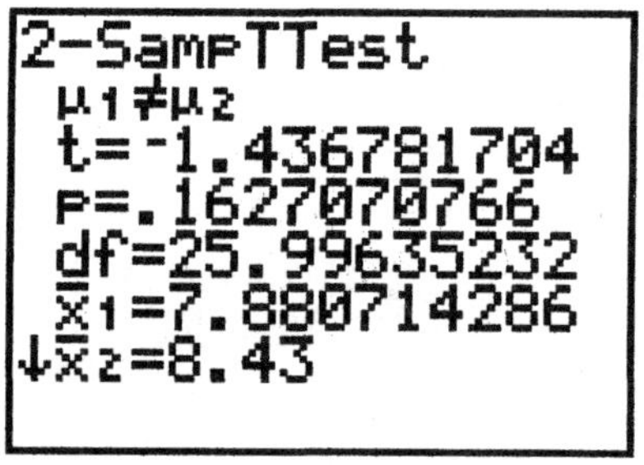

```
2-SampTTest
 μ1≠μ2
 t=-1.436781704
 P=.1627070766
 df=25.99635232
 x1=7.880714286
↓x2=8.43
```

The output (shown above) for **Calculate** displays the alternative hypothesis, the test statistic, the P-value, the adjusted degrees of freedom and the sample statistics.

Or, select **Draw** and press **ENTER**.

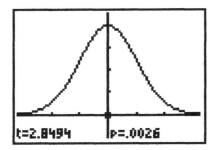

```
t=2.8494        p=.0026
```

Since the P-value is less than α, we reject H_0. The data supports the claim that the treatment group has a larger mean improvement than the control group.

To construct a 95% confidence interval for $(\mu_1 - \mu_2)$, press **STAT**, highlight **TESTS** and select **0:2-SampTint**. For **Inpt,** select **Stats** and press **ENTER**. Enter the sample statistics for each group. Scroll down to the next line and enter .95 for the **C-level**. On the next line, select **No**, because we are not using a pooled variance.

Scroll down to the next line, **Calculate,** and press **ENTER**

```
2-SampTInt
 (2.0412,11.359)
 df=112.4181377
 x̄1=14.8
 x̄2=8.1
 Sx1=12.5
↓Sx2=12.7
■
```

A 95% confidence interval for the difference in the population means is (2.04,11.36). We can interpret this in the following way: the mean improvement in the total score for the treatment group is 2.04 to 11.36 points more than the mean improvement for the control group. (Note: This interval differs slightly from the textbook interval because the calculator uses the adjusted degrees of freedom in the calculations.)

▶ Problem 11 (pg. 616)

To test the claim that carpeted rooms contain more bacteria than uncarpeted rooms, use a one-tailed test: H_o: $\mu_1 = \mu_2$ vs H_a: $\mu_1 > \mu_2$. Enter the data from the 8 carpeted rooms into **L1** and the data from the uncarpeted rooms into **L2**. Because the data sets are small, the first step is to use normal probability plots and boxplots to verify that both datasets are approximately normal and contain no outliers.

The graphs for the carpeted rooms are shown here:

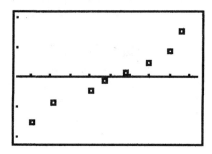

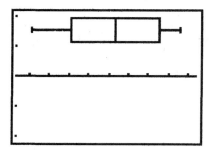

The normal probability plot is fairly linear indicating that the dataset can be assumed to be normally distributed. The boxplot indicates that there are no outliers.

Use the same procedure for the second dataset. You will find that the normal probability plot indicates normality and that the boxplot indicates that there are no outliers.

To run the hypothesis test, press **STAT**, highlight **TESTS**, and select **4:2-SampTTest**. Since you are inputting the sample data, select **Data** and press **ENTER**. Enter **L1** for **List1** and **L2** for **List 2**. Set **Freq1** and **Freq2** to 1. Select >μ_2 as the alternative hypothesis and press **ENTER**. Scroll down to the next line. Select **NO** because, in the procedure we are using, we do not want a pooled variance. Press ENTER. Scroll down to the next line, highlight **Calculate** and press ENTER.

```
2-SampTTest
 μ1>μ2
 t=.9557706856
 p=.1779571293
 df=13.56321553
 x̄1=11.2
↓x̄2=9.7875
```

The output (shown above) for **Calculate** displays the alternative hypothesis, the test statistic, the P-value, the adjusted degrees of freedom and the sample statistics.

If you choose **Draw**, the output includes a graph with the area associated with the P-value shaded.

Since the P-value (.178) is greater than α, we fail to reject H_0. There is not sufficient evidence to support the claim that there is more bacteria in carpeted rooms than in uncarpeted rooms.

To construct a 95% confidence interval for ($\mu_1 - \mu_2$), press STAT, highlight **TESTS** and select **0:2-SampTint**. For **Inpt,** select **Data** and press ENTER. Enter **L1** and **L2** for **List1** and **List2**. Scroll down and enter .95 for the **C-level**. On the next line, select **No**, because we are not using a pooled variance.

Scroll down to the next line, **Calculate,** and press ENTER

```
2-SampTInt
 (-1.767,4.5918)
 df=13.56321553
 x̄1=11.2
 x̄2=9.7875
 Sx1=2.6774188
↓Sx2=3.21000111
```

A 95% confidence interval for the difference in the population means is (-1.8,4.8) Since this interval contains 0, the correct conclusion is that there is no difference in the amount of bacteria in carpeted rooms vs. uncarpeted rooms. (Note: This interval differs slightly from the textbook interval because the calculator uses the adjusted degrees of freedom in the calculations.)

Section 10.3

▶ Example 1 (pg. 622) Testing a Claim Regarding Two
Population Proportions

To test the claim that the proportion of Nasonex users who experienced
headaches as a side effect is greater than the proportion in the control group who
experienced headaches, the correct hypothesis test is: $H_o: p_1 = p_2$ vs $H_a: p_1 > p_2$. Designate the Nasonex users as Group 1 and the Control Group as Group 2.
The sample statistics are $n_1 = 2103$, $x_1 = 547$, $n_2 = 1671$, and $x_2 = 368$.

First, verify that the requirements for the hypothesis test are satisfied. The
problem states that the individuals were randomly divided into two groups so the
first requirement (independent random samples) is satisfied. The second
requirement is: $n\hat{p}(1 - \hat{p}) \geq 10$ for each of the groups. For the first group,

$\hat{p} = \dfrac{x}{n} = \dfrac{547}{2103} = 0.26$. The calculation, $n\hat{p}(1 - \hat{p}) = 2103 * 0.26 * (1 - 0.26)$

is greater than 10. Repeat this calculation for the second group. Both
calculations are greater than 10 so the second requirement is satisfied. The final
requirement is that the sample sizes are not more than 5% of the population sizes.
The population of Americans 12 years of age or older is in the millions so this
requirement is easily satisfied.

Next, to run the hypothesis test, press **STAT**, highlight **TESTS** and select **6:2-
PropZTest** and fill in the appropriate information. Highlight **Calculate** and
press **ENTER**.

```
2-PropZTest
 P1>P2
 z=2.839330068
 P=.0022604818
 P1=.2601046125
 P2=.2202274087
↓P=.2424483307
```

The output displays the alternative hypothesis, the test statistic, the P-value, the
sample proportions, the weighted estimate of the population proportion, $\hat{p}$, and
the sample sizes.

Or, highlight **Draw** and press ENTER.

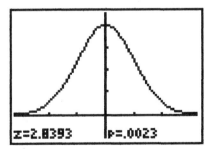

z=2.8393   P=.0023

Since the P-value is less than α, the correct decision is to **Reject** H_o. There is sufficient evidence to support the claim that there is a higher incidence of headaches among the Nasonex users than among the individuals taking a placebo.

◀

| Example 2 (pg. 625) | A Confidence Interval for the Difference between Two Population Proportions |

Construct a confidence interval to compare the proportion of Nasonex users who experience headaches to the proportion of nonusers who experience headaches. Designate the Nasonex users as Group 1 and the Control group as Group 2. The sample statistics are $n_1 = 374$, $x_1 = 64$, $n_2 = 376$, and $x_2 = 68$.

First, verify that the requirements for the hypothesis test are satisfied. The problem states that the individuals were randomly divided into two groups so the first requirement (independent random samples) is satisfied. The second requirement is: $n\hat{p}(1 - \hat{p}) \geq 10$ for each of the groups. For the first group,

$$\hat{p} = \frac{x}{n} = \frac{64}{374} = 0.17.$$ The calculation, $n\hat{p}(1 - \hat{p}) = 374 * 0.17 * (1 - 0.17)$ is

greater than 10. Repeat this calculation for the second group. Both calculations are greater than 10 so the second requirement is satisified. The final requirement is that the sample sizes are not more than 5% of the population sizes. The population of Americans between the ages of 3 and 11 is in the millions so this requirement is easily satisfied.

Press **STAT**, highlight **TESTS** and select **B:2-PropZInt** and fill in the appropriate information. Highlight **Calculate** and press **ENTER**.

```
2-PropZInt
 (-.0555,.03601)
 p̂₁=.1711229947
 p̂₂=.1808510638
 n₁=374
 n₂=376
```

The confidence interval (-.056, .036) contains 0. This means that there is no evidence to support the claim that the proportion of Nasonex patients complaining of headaches is different from those individuals who do not take Nasonex.

◀

Section 10.4

> ▶ Example 2 (pg. 637) Testing a Claim Regarding Two
> Population Standard Deviations

To test the claim that Cisco Systems is a more volatile stock than General Electric, use a one-tailed test: H_o: $\sigma_1 = \sigma_2$ vs H_a: $\sigma_1 > \sigma_2$. Enter the data from the Cisco Systems stock into **L1** and the data from the General Electric stock into **L2**.

The hypothesis test we are using requires that the data be normally distributed. Even minor deviations from normality will affect the validity of the test. (Note: In other hypothesis tests, minor deviations from normality did not seriously affect the test validity.) So, an important first step is to use normal probability plots to verify that both datasets are normally distributed. Once this has been confirmed, we can run the hypothesis test.

To run the hypothesis test, press **STAT**, highlight **TESTS**, and select **D:2-SampFTest**. Since you are inputting the sample data, select **Data** and press **ENTER**. Enter **L1** for **List1** and **L2** for **List 2**. Set **Freq1** and **Freq2** to **1**. On the next line, select $> \sigma_2$ as the alternative hypothesis. Highlight **Calculate** and press **ENTER**.

```
2-SampFTest
 σ1>σ2
 F=7.519296727
 p=6.9363271E-4
 Sx1=11.8356524
 Sx2=4.31622017
↓x̄1=8.333
```

The output displays the alternative hypothesis, the test statistic (F), the P-value, the sample standard deviations, the sample means and the sample sizes.

Or, highlight **Draw** and press **ENTER**.

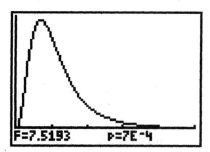

The output shows the F-distribution curve, the value of the F-statistic and the P-value. Since the P-value (.0007) is less than α , the correct decision is to **Reject** H_o . There is sufficient evidence to support the claim that Cisco Systems stock is more volatile than General Electric stock.

▶ Example 3 (pg. 638) Testing a Claim Regarding Two
Population Standard Deviations

To test the claim that the standard deviation of the red blood cell count in the flight animals is different from the standard deviation of the red blood cell count in the control animals , use a two-tailed test: $H_o: \sigma_1 = \sigma_2$ vs. $H_a: \sigma_1 \neq \sigma_2$. Enter the data from the flight rats into **L1** and the data from the control rats into **L2**. Use normal probability plots to verify that each data set is normally distributed.

To run the hypothesis test, press **STAT**, highlight **TESTS**, and select **D:2-SampFTest**. Since you are inputting the sample data, select **Data** and press **ENTER**. Enter L1 for **List1** and L2 for **List 2**. Set **Freq1** and **Freq2** to **1**. On the next line, select $\neq \sigma_2$ as the alternative hypothesis. Highlight **Calculate** and press **ENTER**.

```
2-SampFTest
 σ1≠σ2
 F=1.023974926
 p=.9665798398
 Sx1=1.0174513
 Sx2=1.00546966
↓x̄1=7.880714286
■
```

The output displays the alternative hypothesis, the test statistic (F), the P-value, the sample standard deviations, the sample means and the sample sizes.

Or, highlight **Draw** and press **ENTER**.

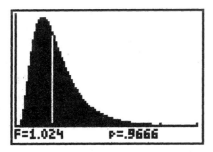

```
F=1.024          p=.9666
```

The output shows the F-distribution curve, the value of the F-statistic and the P-value. Since the P-value (.9666) is greater than α , the correct decision is to **Fail to Reject** H_o. There is not sufficient evidence to support the claim that the standard deviation of the red blood cell count for the flight rats is different from the standard deviation o the red blood cell count for the control group.

Chi-Square Procedures

CHAPTER

11

Section 11.1

▶ Example 2 (pg. 659) Testing a Claim using the Goodness-of-Fit Test

In this example, we test the claim that the population distribution of the United States is different now than it was in 1999. The procedure that we use is the Chi-Square (χ^2) Goodness-of-Fit test.

The χ^2 test has 3 requirements: (1) the data are randomly selected; (2) all *expected frequencies* are greater than or equal to 1 and (3) no more than 20% of the *expected frequencies* are less than 5. For the first requirement, we *assume* that the data was randomly selected. For requirements (2) and (3), we set up a table using the data and then check to see if these requirements are satisfied.

Enter the percentages (in decimal form) given in the problem for each of the four regions into **L1**. Move the cursor so that it is flashing on 'L2' at the top of the second column and press ENTER. The cursor will move to the bottom of the screen and will be flashing next to 'L2='. Type in **L1*2000**. (Note: 2000 is the sample size.) Press ENTER . **L2** will contain the *expected frequencies*.

Enter the *observed frequencies* into **L3**.

L1	L2	L3	3
.196	392	365	
.23	460	404	
.354	708	752	
.22	440	479	
------	------	------	
L3(5) =			

Notice the *expected frequencies* in L2. All the values are greater than 5, so the requirements of the test have been satisfied.

This test is a test of the hypotheses: H_o: The distribution of residents in the U.S. is the same today as it was in 1999, vs. H_a: The distribution of residents in the U.S. is different today than in 1999.

The test statistic is $\chi^2 = \sum \dfrac{(O_i - E_i)^2}{E_i}$. To calculate the χ^2-value, move the cursor so that it is flashing on 'L4' at the top of the fourth column. (Note: To view the fourth column, use the blue right arrow.) Press ENTER . The cursor will move to the bottom of the screen and will be flashing next to 'L4='. Type in (L3-L2)^2/L2. Press ENTER . Press 2nd [QUIT]. Next, press 2nd [LIST]. Select **Math** and **5:sum(** . Type in **L4** and press ENTER. This is the value of the χ^2 statistic.

To calculate the P-value associated with this test statistic of 14.87, we must find the area to the *right* of 14.87 in the χ^2 curve. Press 2nd [DISTR] and select 7: χ^2 cdf. The calculation requires *a lower bound, an upper bound and the degrees of freedom.)* The lower bound is the test statistic, the upper bound is positive infinity (1E99) and *degrees of freedom* is equal to the 'number of categories minus 1'. In this example: df = 4 (regions in the U.S.) −1.

```
X²cdf(14.87,1E99
,3)
        .0019311812
```

Since the P-value (.002) is less than α (.05), the correct decision is to **reject** the null hypothesis. There is sufficient evidence to support the claim that the distribution of residents in the U.S. is different today than it was in 1999.

◀

▶ Example 3 (pg. 660)   Testing a Claim Using the Goodness
                          Of Fit Test

In this example, we test the claim that the day (Sunday, Monday, Tuesday, etc.) on which a child is born occurs with equal frequency. This is a test of the

hypotheses: H_o: $p_1 = p_2 = p_3 = p_4 = p_5 = p_6 = p_7 = \dfrac{1}{7}$ vs. H_a: At least one

of the proportions is different than the others.

Enter the fraction $\dfrac{1}{7}$ into **L1** seven times (for the seven days of the week). Move

the cursor so that it is flashing on 'L2' at the top of the second column and press ENTER. The cursor will move to the bottom of the screen and will be flashing next to 'L2='. Type in **L1*500**. (Note: 500 is the sample size.) Press ENTER. **L2** will contain the *expected frequencies*.

Enter the *observed frequencies* into **L3**.

L1	L2	L3	3
.14286	71.429	78	
.14286	71.429	74	
.14286	71.429	76	
.14286	71.429	71	
.14286	71.429	81	
.14286	71.429	63	
------	------	------	

L3(8) =

Notice the *expected frequencies* in L2. All the values are greater than 5, so the requirements of the test have been satisfied.

The test statistic is $\chi^2 = \sum \dfrac{(O_i - E_i)^2}{E_i}$. To calculate the χ^2-value, move the

cursor so that it is flashing on 'L4' at the top of the fourth column. (Note: To view the fourth column, use the blue right arrow.) Press ENTER. The cursor will move to the bottom of the screen and will be flashing next to 'L4='. Type in **(L3-L2)²/L2**. Press ENTER. Press 2nd **[QUIT]**. Next, press 2nd **[LIST]**. Select **Math** and **5:sum(**. Type in **L4** and press ENTER. This is the value of the χ^2 statistic.

To calculate the P-value associated with this test statistic of 6.184, we must find the area to the *right* of 6.184 in the χ^2 curve. Press 2ⁿᵈ [DISTR] and select 7: χ^2 cdf. The calculation requires *a lower bound, an upper bound and the degrees of freedom.)* The lower bound is the test statistic, the upper bound is positive infinity (1E99) and *degrees of freedom* is equal to the 'number of categories minus 1'. In this example: df = 7 (days of the week) –1.

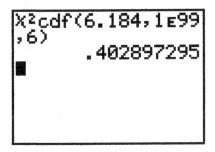

Since the P-value (.403) is greater than α (.01), the correct decision is to **fail to reject** the null hypothesis. There is not sufficient evidence to support the claim that the day of the week on which a child is born occurs with equal frequency.

▶ Example 4 (pg. 662)   Testing a Claim that a Random Variable
Follows a Binomial Distribution

In this example, we are testing whether a set of data follows a binomial distribution. This is a test of the hypotheses: H_o: The random variable, X, is a binomial random variable with n = 4 and p = 0.6, vs. H_a: The random variable, X, is **not** a binomial random variable with n = 4 and p = 0.6.

In this experiment, the basketball player shoots four free throws and records the number of shots that she makes. The experiment is repeated 200 times.

Enter the values 0,1,2,3,4 into **L1**. These are the four possible values of the random variable. Press 2nd [QUIT]. Press 2nd [DISTR] and select **0: binompdf**. Enter the values for n and p, and press ENTER. The probability values associated with the four values of the random variable will be displayed. Press STO **L2** and press ENTER . The values will be stored into **L2**.

```
binompdf(4,.6
{.0256 .1536 .3…
Ans→L₂█
```

Press STAT , select **EDIT**. **L1** contains the values of the random variable, X, and **L2** contains their probabilities, assuming a binomial model with n=4 and p=0.6. Move the cursor so that it is flashing on 'L3' at the top of the third column and press ENTER. The cursor will move to the bottom of the screen and will be flashing next to 'L3='. Type in **L2*200**. (Note: 200 is the sample size.) Press ENTER . **L3** will contain the *expected frequencies*.

Use the blue right arrow to move to **L4** and enter the *observed frequencies*

L2	L3	L4	4
.0256	5.12	10	
.1536	30.72	52	
.3456	69.12	67	
.3456	69.12	56	
.1296	25.92	15	
------	------	------	
L4(6) =			

Notice the *expected frequencies* in L3. All the values are greater than 5, so the requirements of the test have been satisfied.

The test statistic is $\chi^2 = \sum \dfrac{(O_i - E_i)^2}{E_i}$. To calculate the χ^2-value, move the cursor so that it is flashing on 'L5' at the top of the fifth column. (Note: To view the fifth column, use the blue right arrow.) Press ENTER. The cursor will move to the bottom of the screen and be flashing next to 'L5='. Type in (L4-L3)^2/L3. Press ENTER. Press 2nd [QUIT]. Next, press 2nd [LIST]. Select **Math** and **5:sum(** . Type in **L5** and press ENTER. This is the value of the χ^2 statistic.

To calculate the P-value associated with this test statistic of 26.55, we must find the area to the *right* of 26.55 in the χ^2 curve. Press 2nd [DISTR] and select 7: χ^2 cdf. The calculation requires *a lower bound, an upper bound and the degrees of freedom.)* The lower bound is the test statistic, the upper bound is positive infinity (1E99) and *degrees of freedom* is equal to the 'number of categories minus 1'. In this example: df = 5 (the number of different possible outcomes of the random variable) $-$ 1 = 4.

```
X²cdf(26.55,1E99
,4)
   2.450850778E-5
```

Since the P-value (.0000245) is less than α (.05), the correct decision is to **reject** the null hypothesis. There is sufficient evidence to reject the claim that the player's free throws follow a binomial distribution with n = 4 and p = 0.6.

▶ Problem 15 (pg.666)

(a.) In this problem, we will use the TI-83 to generate 500 random integers numbered 1 through 5. The first step is to set the *seed* by selecting a 'starting number' and storing this number in **rand**. Suppose, for this example, that we select the number '22' as the starting number. Type **22** into your calculator and press the ▮STO▮ key. Next press the ▮MATH▮ key and select **PRB** and select **rand,** which stands for 'random number'. Press ▮ENTER▮ and the starting value of '22' will be stored into **rand** and will be used as the *seed* for generating random numbers.

Now you are ready to generate the random integers. Press ▮MATH▮ and select **PRB**. Select **5:RandInt(**. This function requires three values: the starting value, the ending value and the number of values you want to generate. For this example, you want to generate 500 values from the integers ranging from 1 to 5. The command is **randInt(1,5,500)**. Press ▮STO▮ **L1** and press ▮ENTER▮ . The values will be stored into **L1**.

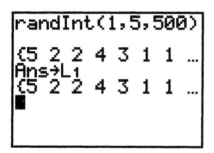

Next, press 2nd [LIST]. Select **OPS** and **1:SortA(** . Type in **L1** and press ▮ENTER▮. This will put all the values in **L1** in numerical order. Press 2nd [QUIT]. Press ▮STAT▮ , select **EDIT.** Move the cursor to the first entry in L1 and scroll through the list using the blue down arrow. Hold the 'down arrow' button down and scroll through the 1's. Record the number of 1's that you have on a piece of paper.

In this illustration, there are 105 1's. (Note: If you started with a different seed, you would get a different result).

Continue scrolling through the column. Stop at the final '2' in the list.

```
┌─────────────────────────────────┐
│ L1      L2      L3         1     │
│ 2                                │
│ 2                                │
│ 2                                │
│ 2                                │
│ 2                                │
│ 2                                │
│ ▓                                │
│─────────                         │
│ L1(205) =3                       │
└─────────────────────────────────┘
```

The number of 2's in this illustration is $204 - 105 = 99$. Record this number on your paper.

Continue this process until you have recorded the counts for all the data values in L1.

(b.) Each of the numbers, 1 through 5, should occur with equal frequency. So, the proportion of 1's, 2's, 3's, 4's and 5's should equal .20.

(c.) In this example, we are testing the claim that the random number generator is generating random numbers between 1 and 5. This is a test of the hypotheses: H_o: $p_1 = p_2 = p_3 = p_4 = p_5 = .2$, vs. H_a: At least one of the proportions is different than the others.

Press **STAT** , select **EDIT**. L1 contains the 500 randomly selected integers. Enter the values 1 through 5 into **L2** and their corresponding frequencies into **L3**. L3 contains the *observed frequencies*

```
┌─────────────────────────────────┐
│ L1      L2      L3         3     │
│ 1       1       105              │
│ 1       2       98               │
│ 1       3       105              │
│ 1       4       88               │
│ 1       5       103              │
│ 1       ------   ------          │
│ 1                                │
│─────────                         │
│ L3(6)  =                         │
└─────────────────────────────────┘
```

Move to **L4** and enter the expected proportion, .20, five times. Move the cursor so that it is flashing on 'L5' at the top of the fifth column and press ENTER. The cursor will move to the bottom of the screen and will be flashing next to 'L5='. Type in **L4*500**. (Note: 500 is the sample size.) Press ENTER. **L5** will contain the *expected frequencies*. All the *expected frequencies* in L5 are greater than 5, so the requirements of the test have been satisfied.

The test statistic is $\chi^2 = \sum \dfrac{(O_i - E_i)^2}{E_i}$. To calculate the χ^2-value, move the cursor so that it is flashing on 'L6' at the top of the sixth column. Press ENTER. The cursor will move to the bottom of the screen and be flashing next to 'L6='. Type in **(L3-L5)²/L5**. Press ENTER. Press 2nd [QUIT]. Next, press 2nd [LIST]. Select **Math** and **5:sum(**. Type in **L6** and press ENTER. This is the value of the χ^2 statistic.

To calculate the P-value associated with this test statistic of 2.07 (in this illustration), we must find the area to the *right* of 2.07 in the χ^2 curve. Press 2nd [DISTR] and select 7: χ^2 cdf. The calculation requires *a lower bound, an upper bound and the degrees of freedom*. The lower bound is the test statistic, the upper bound is positive infinity (1E99) and *degrees of freedom* is equal to the 'number of categories minus 1'. In this example: df = 5 (the number of different possible outcomes of the random variable) – 1 = 4.

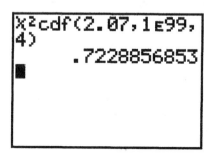

Since the P-value (.723) is greater than α (.01), the correct decision is to **fail to reject** the null hypothesis. There is **not** sufficient evidence to reject the claim that the random number generator is generating random numbers between 1 and 5. In other words, we can conclude that the random number generator is working correctly.

Section 11.3

▶ Example 2 (pg. 681) Performing a Chi-Square Independence Test

In this example we are testing the claim that blood type and Rh level are independent. The correct procedure is the Chi-Square (χ^2) Independence test. The hypotheses are: H_o: Blood type and Rh factor are independent (or not related), vs. H_a: Blood type and Rh factor are dependent (or somehow related).

The χ^2 test has 3 requirements: (1) the data are randomly selected; (2) all *expected frequencies* are greater than or equal to 1 and (3) no more than 20% of the *expected frequencies* are less than 5. For the first requirement, we *assume* that the data was randomly selected. We will verify requirements (2) and (3) at the end of our analysis.

The first step is to enter the data in the table into **Matrix A**. On the TI-83 Plus, press 2ⁿᵈ **[MATRX]** . (On the TI-83, press MATRX). Highlight **EDIT** and press ENTER.

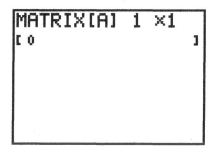

On the top row of the display, enter the size of the matrix. The matrix has 2 rows and 4 columns, so press 2 , press the right arrow key, and press 4. Press ENTER. Enter the first value, 176, and press ENTER. Enter the second value, 28, and press ENTER. Continue this process and fill the matrix.

Press 2nd [Quit] . To perform the test of independence, press STAT, highlight
TESTS, and select **C**: χ^2-**Test** and press ENTER.

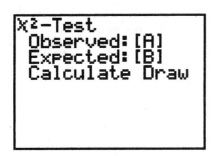

For **Observed**, **[A]** should be selected. If **[A]** is not already selected, press 2nd
[MATRX], highlight **NAMES**, select **1:[A]** and press ENTER. For, **Expected**,
[B] should be selected. Move the cursor to the next line and select **Calculate** and
press ENTER.

The output displays the test statistic and the P-value. Since the P-value is greater
than α, the correct decision is to **Fail to Reject** the null hypothesis. This means
that blood type is *independent* Rh factor.

Or, you could highlight **Draw** and press ENTER.

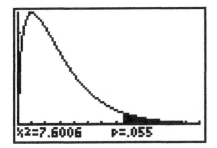

This output displays the χ^2 –**curve** with the area associated with the P-value shaded in. The test statistic and the P-value are also displayed.

The final step in this procedure is to confirm that the test requirements have been satisfied. The two requirements that we need to verify are
: (1) all *expected frequencies* are greater than or equal to 1 and (2) no more than 20% of the *expected frequencies* are less than 5. Both requirements involve the *expected frequencies* which are stored in **Matrix B**. To view **Matrix B,** press 2nd **[MATRX]** highlight **NAMES**, select **2:[B]** and press ENTER ENTER.

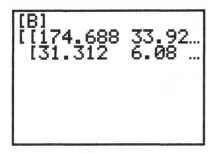

Scroll through the 8 entries in **Matrix B** and confirm that all entries are greater than 5. This confirms that the test requirements have been satisfied.

▶ Example 4 (pg. 685)   A Test of Homogeneity of Proportions

In this example, we are testing the claim that the proportions of subjects who experience abdominal pain are equal among all three groups (those taking Zocor, those taking a placebo and those taking Cholestyramine). The correct procedure is the Chi-Square (χ^2) Independence test. The hypotheses are: H_o :

$p_1 = p_2 = p_3$ vs. H_a: At least one of the proportions is different than the others.

The χ^2 test has 3 requirements: (1) the data are randomly selected; (2) all *expected frequencies* are greater than or equal to 1 and (3) no more than 20% of the *expected frequencies* are less than 5. For the first requirement, we *assume* that the data was randomly selected. We will verify requirements (2) and (3) at the end of our analysis.

The first step is to enter the data in the table into **Matrix A**. On the TI-83 Plus, press 2nd [MATRX]. MATRX is found above the ■ key. (On the TI-83, press MATRX). Highlight **EDIT** and press ENTER.

On the top row of the display, enter the size of the matrix. The matrix has 2 rows and 3 columns, so press 2 , press the right arrow key, and press ■ Press ENTER. Enter the first value, 51, and press ENTER. Enter the second value, 5, and press ENTER. Continue this process and fill the matrix.

Press 2nd [Quit] . To perform the test of independence, press STAT, highlight **TESTS**, and select **C: χ^2-Test** and press ENTER.

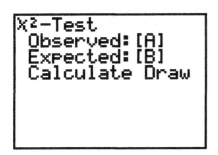

For **Observed**, [A] should be selected. If [A] is not already selected, press 2nd [MATRX], highlight **NAMES**, select **1:[A]** and press ENTER. For, **Expected**, [B] should be selected. Move the cursor to the next line and select **Calculate** and press ENTER.

```
X²-Test
 X²=14.70651321
 P=6.4050309E⁻4
 df=2

■
```

The output displays the test statistic and the P-value. Since the P-value is less than α, the correct decision is to **Reject** the null hypothesis. We conclude that at least one of the three groups experiences abdominal pain at a rate different from the other two groups.

The final step in this procedure is to confirm that the test requirements have been satisfied. The two requirements that we need to verify are: (1) all *expected frequencies* are greater than or equal to 1 and (2) no more than 20% of the *expected frequencies* are less than 5. Both requirements involve the *expected frequencies* which are stored in **Matrix B**. To view **Matrix B,** press 2nd [MATRX] highlight **NAMES**, select **2:[B]** and press ENTER ENTER.

```
 df=2


[B]
[[59.39343408  5…
 [1523.606566  1…
■
```

Scroll through the 6 entries in **Matrix B** and confirm that all entries are greater than 5. This confirms that the test requirements have been satisfied.

◀

▶ Problem 7 (pg. 689)

In this example, we are testing the claim that education level is independent of region of the United States. The correct procedure is the Chi-Square (χ^2) Independence test. The hypotheses are: H_o : *education level* is independent of *region of the United States*. vs. H$_a$: *education level* is not independent of *region of the United States*.

Enter the data in the table into **Matrix A**.. Highlight **EDIT** and press ENTER. On the top row of the display, enter the size of the matrix. The matrix has 4 rows and 4 columns. Press ENTER. Fill the matrix with the data values.

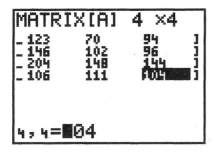

Press 2nd [Quit] . To perform the test of independence, press STAT, highlight **TESTS**, and select **C: χ^2-Test** and press ENTER.

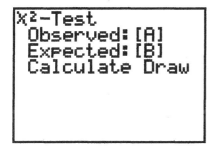

For **Observed**, **[A]** should be selected. If **[A]** is not already selected, press 2nd **[MATRX]**, highlight **NAMES**, select **1:[A]** and press ENTER. For, **Expected**, **[B]** should be selected. Move the cursor to the next line and select **Calculate** and press ENTER.

```
X²-Test
 X²=32.92570338
 P=1.3758851ᴇ-4
 df=9



■
```

The output displays the test statistic and the P-value. Since the P-value is less than α, the correct decision is to **Reject** the null hypothesis. We conclude that *education level* and *region of the country* are **not** independent.

The final step in this procedure is to confirm that the test requirements have been satisfied. The two requirements that we need to verify are: (1) all *expected frequencies* are greater than or equal to 1 and (2) no more than 20% of the *expected frequencies* are less than 5. Both requirements involve the *expected frequencies* which are stored in **Matrix B**. To view **Matrix B**, press 2nd **[MATRX]** highlight **NAMES**, select **2:[B]** and press ENTER ENTER.

```
[B]
[[66.89800443  1…
 [92.15742794  1…
 [121.3636364  1…
 [75.58093126  1…
```

Scroll through the 16 entries in **Matrix B** (as you scroll through the entries, record these entries (rounded to the nearest whole number) on paper to use in part (b). of these problem) and confirm that all entries are greater than 5. This confirms that the test requirements have been satisfied.

(b). To determine which cell contributed most to the test statistic, we will store the two matrices into Lists.

Press STAT, highlight **EDIT.** Enter the *observed values* into L1 and enter the *expected values* into L2. To calculate each cell's contribution to the χ^2-value, move the cursor so that it is flashing on 'L3' at the top of the third column. Press ENTER . The cursor will move to the bottom of the screen and will be flashing next to 'L3='. Type in $(L1-L2)^2/L2$. Press ENTER . Scroll through the values in L3 and find the largest value, 3.9. This is the entry for 'West' region and 'Some College.' The observed value for this cell is '111' and the expected value is '93'. This tells us that, for this cell, **more** residents had 'some college' than is expected.

◀

Inferences on the Least Squares Regression Model; ANOVA

Section 12.1

▶ Example 1 (pg. 706) Least-Squares Regression

Press **STAT**, highlight **1:Edit** and clear **L1** and **L2**. For each of the fourteen patients, enter the age into **L1** and the total cholesterol into **L2**. Press **2ⁿᵈ** **[STAT PLOT]** , select **1:Plot1**, turn **ON** Plot 1 and press **ENTER**. For **Type** of graph, select the **scatter plot** which is the first selection. Press **ENTER**. Enter **L1** for **Xlist** and **L2** for **Ylist**. Highlight the first selection, the small square, for the type of **Mark**. Press **ENTER**. Press **ZOOM** and **9** to select **ZoomStat**.

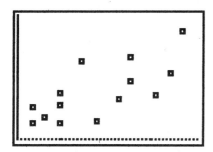

This graph shows a positive linear correlation, with quite a bit of scatter.

In order to calculate r, the correlation coefficient, and, r^2, the coefficient of determination, you must turn **On** the **Diagnostic** command. Press **2ⁿᵈ** **[CATALOG]** (Note: **CATALOG** is found above the **0** key). The CATALOG of functions will appear on the screen. Use the down arrow to scroll to the **DiagnosticOn** command.

Press **ENTER** **ENTER**.

To calculate the correlation coefficient, the coefficient of determination and the regression equation, press **STAT**, highlight **CALC**, select **4:LinReg(ax+b)**. Press **VARS**, highlight **Y-VARS**, select **1:Function** by pressing **ENTER**, select **1:Y1** by pressing **ENTER** and press **ENTER**. This stores the regression equation in **Y1**. (Note: This command requires that you specify which lists contain the X-values and Y-values. If you do not specify these lists, the defaults are used. The defaults are: **L1** for the X-values and **L2** for the Y-values.)

```
LinReg
 y=ax+b
 a=1.399064152
 b=151.3536582
 r²=.5152520915
 r=.7178106237
█
```

The correlation coefficient is r = .718. This suggests a positive linear correlation between X and Y, but not a very strong one. The coefficient of determination is .515. This tells us that 51.5% of the variation in cholesterol levels can be explained by the predictor variable, age. The regression equation is
$\hat{y} = 151.35 + 1.399x$.

To see a scatterplot of the data along with the regression equation, press **GRAPH**

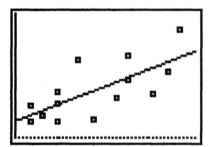

Example 2 (pg. 710)   Computing the Standard Error

Method 1: Using this method, we will go through all the steps for calculating the standard error. We will actually generate the columns of 'residuals' and 'residuals^2' as they appear in Table 2 on pg. 710.

Enter the data from Table 1 on pg. 706 into **L1** and **L2**. Press **STAT**, highlight **CALC**, select **4:LinReg(ax+b)**. Press **VARS**, highlight **Y-VARS**, select **1:Function** by pressing **ENTER**, select **1:Y1** by pressing **ENTER** and press **ENTER**. This stores the regression equation in **Y1**.

The formula for s_e, the standard error of the estimate is $\sqrt{\dfrac{\sum (y_i - \hat{y}_i)^2}{n-2}}$. The values for $(y_i - \hat{y}_i)$, called Residuals, are automatically stored to a list called **RESID**. Press **2nd** **[LIST]**, select **7:RESID**. Press **STO**, **2nd** **[L3]** and **ENTER**. This stores the residuals to **L3**.

```
LRESID→L₃
{-6.33026198 8.…
■
```

In the formula for s_e, the residuals, $(y_i - \hat{y}_i)$, are squared. To square these values and store them in **L4**, press **STAT**, select **1:Edit** and move the cursor to highlight the Listname **L4**. Press **ENTER**. Press **2nd** **[L3]** and the x^2 key.

L2	L3	L4	4
180	-6.33	------	
195	8.6697		
186	-4.527		
180	-16.12		
210	13.876		
197	.87629		
239	34.482		

L4 =L₃^2

Press **ENTER**.

```
 L2       L3       L4       4
 180      -6.33    █████████
 195      8.6697   75.164
 186      -4.527   20.498
 180      -16.12   259.97
 210      13.876   192.55
 197      .87629   .76788
 239      34.482   1189
 L4(1)=40.07221673…
```

Press **2ⁿᵈ** **[QUIT]** . Next, you need to find the sum of **L4**, $\sum (y_i - \hat{y}_i)^2$. Press **2ⁿᵈ** **[LIST]** . Highlight **MATH**, select **5:sum(** and press **2ⁿᵈ** **[L4]** . Close the parentheses and press **ENTER**.

```
sum(L4)
            4553.894977
█
```

Divide this sum by (number of observations – 2). In this example, (n - 2) = 12. Simply press ÷ 12 and press **ENTER**.

```
sum(L4)
            4553.894977
Ans/12
            379.4912481
```

Lastly, take the square root of this answer by pressing **2ⁿᵈ** **[√]** **2ⁿᵈ** **[ANS]** . Close the parentheses and press **ENTER**.

```
sum(L4)
          4553.894977
Ans/12
          379.4912481
√(Ans)
          19.48053511
■
```

The standard error of the estimate, s_e, is 19.48.

Method 2: Using this method, the standard error can be obtained directly, without going through the individual steps in the calculation.

Press **STAT**, highlight **TESTS** and select **E:LinRegTTest**. Enter **L1** for **Xlist**, **L2** for **Ylist**, and **1** for **Freq**. On the next line, β and ρ, select $\neq 0$ and press **ENTER**. Leave the next line, RegEQ, blank. Highlight **Calculate**.

```
LinRegTTest
 Xlist:L1
 Ylist:L2
 Freq:1
 β & ρ:≠0 <0 >0
 RegEQ:
 Calculate
```

Press **ENTER**.

```
LinRegTTest
 y=a+bx
 β>0 and ρ>0
 t=3.57143321
 p=.0019211307
 df=12
↓a=151.3536582
```

The output displays several pieces of information describing the relationship between X and Y. What you are interested in for this example is the standard error. Scroll down to the next page of output and you will see s = 19.48. This is the standard error.

◀

▶ Example 3 (pg. 711)   Verifying that the Residuals are Normally
                         Distributed

This example is a continuation of Example 2. In Example 2, the Residuals
obtained from the regression procedure were stored in **L3**. To set up the normal
probability plot, press **2ⁿᵈ [STAT PLOT]** . Press **ENTER**
to select **Plot 1**. Highlight **On** and press **ENTER**. Set **Type** to the normal
probability plot which is the third selection in the second row. Press **ENTER**.
Set **Data List** to **L3** and **Data Axis** to **X**. For **Marks** select the small square.

Press **ZOOM** and select **9:ZoomStat** and **ENTER**.

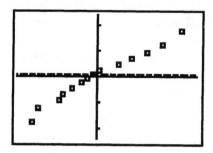

This plot is *fairly* linear, indicating that the Residuals follow a normal
distribution.

▶ Example 4 (pg. 713)   Testing for a Linear Relation

This example is a continuation of the previous examples (#1-#3).

To test the claim that there is a linear relationship between *age* and *total cholesterol*, the appropriate hypothesis test is: $\beta_1 = 0$ vs. $\beta_1 \neq 0$.

The first step is to verify that the assumptions required to perform the test are satisfied. The first assumption is that the sample has been obtained using random sampling. This has been confirmed and is stated in Example 1. The next assumption is that the residuals are normally distributed and this has been confirmed by the normal probability plot in Example 3. The last assumption is that the residuals have constant error variance. This assumption can be validated using a graph of the Residuals vs. the predictor variable, Age. If the Residuals do, in fact, have constant error variance, then the Residuals will appear as a scatter of points about the horizontal line at 0.

To construct the graph of the Residuals vs. Age, press **2ⁿᵈ [STAT PLOT]** . Press **ENTER** to select **Plot 1**. Highlight **On** and press **ENTER**. Set **Type** to the scatter plot which is the first selection in the first row. Press **ENTER**. Set **Xlist** to **L1** and **YList** to **L3**. For **Marks**, select the small square.

Press **WINDOW** and use the data to set the Window. First, look at the *Age* variable in Table 1 on pg. 706. Notice the minimum and maximum values (25 and 65). Choose **Xmin** and **Xmax** to encompass these values. For example, choose **Xmin** = 15 and **Xmax** = 70. To set the Y-values, you must look through the Residuals that are stored in **L3**. Press **2ⁿᵈ [QUIT]** . Press **STAT**, highlight **EDIT** and scroll through the values in **L3**. The minimum value is –24.5 and the maximum value is 34.482. To make the graph symmetric about the X-axis, use **Ymin** = -35 and **Ymax** = 35. Press **WINDOW** and enter these values for **Ymin** and **Ymax**. Press **GRAPH**

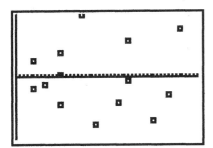

 The errors are evenly spread about the horizontal line at 0, so the assumption of constant error variance is satisfied.

To run the test, press STAT, highlight **TESTS** and select **E:LinRegTTest**. Enter
L1 for **Xlist**, **L2** for **Ylist**, and **1** for **Freq**. On the next line, β and ρ, select
$\ne 0$ and press ENTER. Leave the next line, RegEQ, blank. Highlight **Calculate**.

```
LinRegTTest
 Xlist:L1
 Ylist:L2
 Freq:1
 β & ρ:≠0 <0 >0
 RegEQ:
 Calculate
```

Press ENTER.

```
LinRegTTest
 y=a+bx
 β>0 and ρ>0
 t=3.57143321
 p=.0019211307
 df=12
↓a=151.3536582
```

The output displays several pieces of information describing the relationship
between X and Y. What you are interested in for this example are the following:
the test statistic (t = 3.5714), the P-value (p = .0019). Since the P-value is less
than α, the correct decision is to **Reject** the null hypothesis. This indicates that
there is a linear relationship between X and Y.

▶ Example 5 (pg. 716) Constructing a Confidence Interval about
the Slope of the True Regression Line

This example is a continuation of the previous examples (# 1 - #4).

The 95% confidence interval for β_1, the slope of the true regression line is given
by the following formula:

$$ b_1 \pm t_{\frac{\alpha}{2}} \cdot \frac{s_e}{\sqrt{\sum(x_i - \bar{x})^2}}. $$

From the previous examples in this Section, we have already obtained the
following values: b_1 and s_e. b_1 is the coefficient of x in the regression equation
and s_e is the standard error. The value for $t_{\frac{\alpha}{2}}$ can be found in the *t-table*, in the
$t_{.025}$ column with (n-2) degrees of freedom.

The only part of the formula that we need to calculate is the denominator and we
can use an equivalent form of this expression that can be easily evaluated on the
calculator. The denominator, $\sqrt{\sum(x_i - \bar{x})^2}$, can be written equivalently as:

$\sqrt{\left(\sum x_i^2 - \frac{\left(\sum x_i\right)^2}{n}\right)}$. The values for $\sum x_i$ and $\sum x_i^2$ are stored in your
calculator as part of the regression procedure. To access them, press VARS and
select **5:Statistics**. On the Statistics Menu, select Σ. Then select the first entry,
Σx by pressing ENTER and press ENTER The result, 589, will appear on the
screen. To obtain $\sum x_i^2$, press VARS and select **5:Statistics**. On the Statistics
Menu, select Σ. Then select the second entry, Σx^2, by pressing ENTER and
press ENTER The result is 27253. Enter these values into the equation:

$$ \sqrt{\left(27253 - \frac{(589)^2}{14}\right)} = 49.73. $$

Now fill the values into the formula for the confidence interval. The lower
bound of the 95% confidence interval is: $1.399 - 2.179 \cdot \frac{19.48}{49.73} = 0.545.$
The upper bound is:

$$ 1.399 + 2.179 \cdot \frac{19.48}{49.73} = 2.253. $$

▶ Problem 11 (pg. 720)

In this exercise, we will go directly to the **LinRegTTest**. This procedure combines all the steps for analyzing the model. It has more output for analyzing the regression model then the command **LinReg(ax+b).**

To begin the analysis, enter the data into L1 and L2. Next, press STAT, highlight **TESTS** and select **E:LinRegTTest**. Enter **L1** for **Xlist**, **L2** for **Ylist**, and **1** for **Freq**. On the next line, β and ρ, select $\neq 0$ and press ENTER. Move the cursor to the next line, RegEQ. On this line, tell the calculator where to store the regression equation. We will store it in **Y1**. Press VARS, highlight **Y-VARS**, select **1:Function** by pressing ENTER, select **1:Y1** by pressing ENTER. Highlight **Calculate**.

```
LinRegTTest
 Xlist:L₁
 Ylist:L₂
 Freq:1
 β & ρ:≠0  <0  >0
 RegEQ:Y₁
 Calculate
```

Press ENTER.

```
LinRegTTest
 y=a+bx
 β≠0 and ρ≠0
 t=2.612080589
 p=.0241676528
 df=11
↓a=2.03358388
```

The second page of output contains additional information.

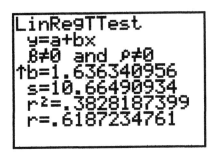

(a.) First, notice the regression equation: y= a+bx. The unbiased estimator of β_0 is 'a' which is equal to 2.034. The unbiased estimator of β_1 is 'b' which is equal to 1.636.

(b.) The standard error is 's' which is equal to 10.665.

(c.) The values for $(y_i - \hat{y}_i)$, called Residuals, are automatically stored to a list called **RESID**. Press **2**nd [LIST] , select **7:RESID**. Press STO, **2**nd [L3] and ENTER. This stores the residuals to **L3**.

To set up the normal probability plot, press **2**nd [STAT PLOT] . Press ENTER to select **Plot 1**. Highlight **On** and press ENTER. Set **Type** to the normal probability plot which is the third selection in the second row. Press ENTER. Set **Data List** to **L3** and **Data Axis** to **X**. For **Marks** select the small square.

Press ZOOM and select **9:ZoomStat** and ENTER.

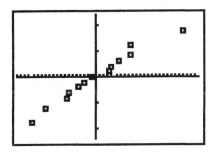

This plot is *fairly* linear, indicating that the Residuals follow a normal distribution. You should, however, notice that there is one unusual value that may be a cause for concern in your analysis. If you press TRACE and use the blue right arrow key, you can scroll through the data and find the actual data point that you find questionable.

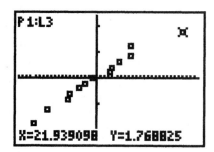

(d.) To calculate s_{b_1}, we will use the formula $\dfrac{s_e}{\sqrt{\sum (x_i - \bar{x})^2}}$. The

denominator, $\sqrt{\sum (x_i - \bar{x})^2}$, can be written equivalently as:

$\sqrt{\left(\sum x_i^2 - \dfrac{(\sum x_i)^2}{n}\right)}$. The values for $\sum x_i$ and $\sum x_i^2$ are stored in your

calculator as part of the regression procedure. To access them, press **VARS** and
select **5:Statistics.** On the Statistics Menu, select Σ. Then select the first entry,
Σx by pressing **ENTER** and press **ENTER** The result, 10, will appear on the
screen. To obtain $\sum x_i^2$, press **VARS** and select **5:Statistics.** On the Statistics
Menu, select Σ. Then select the second entry, Σx², by pressing **ENTER** and
press **ENTER** The result is 292.52. Enter these values into the equation:

$$\dfrac{10.665}{\sqrt{292.52 - \dfrac{(10)^2}{11}}} = .633.$$

(e.) To test the claim that there is a linear relationship between the predictor
variable, 'Rate of return of S&P 500' and 'Rate of return of Cisco Systems', the
appropriate hypothesis test is: $\beta_1 = 0$ vs. $\beta_1 \neq 0$. This is the test that we set up
in the **LinRegTTest**. Notice that a p-value is displayed in the output. Since the
p-value of .02 is less than the α-value of .10, the correct decision is to **reject H₀.**
The data supports the claim that there is a linear relationship between the
variables.

(f.) The 99% confidence interval for β_1 is: $b_1 \pm t_{\frac{\alpha}{2}} s_{b_1}$. Replace all the variables with

their appropriate values and obtain: 1.636±3.106(.633).

(g.) To obtain the mean rate of return for Cisco Systems stock if the rate of return of the
S&P 500 is 4.2 percent, press **VARS,** highlight **Y-VARS,** select **1:Function** by pressing
ENTER , select **1:Y1** by pressing **ENTER** Type in (4.2) and press **ENTER**
Y1(4.2)=8.91.

Section 12.2

▶ **Example 1 (pg. 725)** Constructing a Confidence Interval about the Mean Predicted Value

This example is a continuation of Examples 1 – 5 in Section 12.1

To construct a confidence interval for the mean y-value at a specific x-value, we must obtain the value for $\hat{y}$ from the regression equation and then calculate the margin of error. The formula that we will use for the margin of error is equivalent to the one in the textbook. The formula we will use is:

$$E = t_{\frac{\alpha}{2}} s_e \sqrt{(\frac{1}{n} + \frac{n(x^* - \bar{x})^2}{n(\sum x_i^2) - (\sum x_i)^2}})$$

The regression equation that we obtained in Example 1 is: $\hat{y} = 151.35 + 1.399x$. Using this equation we calculate $\hat{y}$ for x=42 and obtain the value, 210.1.

The critical value for t is found in the t-table. For this example, $t_{\frac{\alpha}{2}}$ with (n-2) degrees of freedom = 2.179.

The standard error, s_e, which we obtained in Example 2 is 19.48.

To calculate $\bar{x}$, $\sum x^2$, and $(\sum x)^2$, press **VARS**, select **5:Statistics**. Highlight **2:** $\bar{x}$ and press **ENTER ENTER**. Notice that $\bar{x} = 42.07$. Press **VARS** again, select **5:Statistics**, highlight $\sum$, select **1:** $\sum x$ and press **ENTER ENTER**. So, $\sum x = 589$. Press **VARS** again, select **5:Statistics**, highlight $\sum$, select **2:** $\sum x^2$, and press **ENTER ENTER**. Notice that $\sum x^2 = 27253$.

Now, calculate the margin of error, E, when $x^* = 42$ using the formula for E shown above.

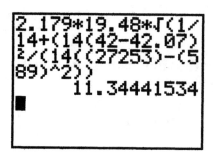

The confidence interval for the mean y-value at x = 42 is $\hat{y} \pm 11.34$. Since $\hat{y} = 210.1$, we calculate 210.1 ± 11.34 to obtain the lower bound of 198.76 and the upper bound of 221.44. These are the lower and upper bounds of the 95% confidence interval for the mean cholesterol level of all 42-year-old females.

> ▶ Example 2 (pg. 726) Constructing a Prediction Interval about
> a Predicted Value

This example is a continuation of Examples 1 – 5 in Section 12.1

To construct a prediction interval for the y-value at a specific x-value, we must obtain the value for $\hat{y}$ from the regression equation and then calculate the margin of error. The formula that we will use for the margin of error is equivalent to the one in the textbook.   The formula we will use is:

$$E = t_{\frac{\alpha}{2}} s_e \sqrt{(1 + \frac{1}{n} + \frac{n(x^* - \bar{x})^2}{n(\sum x_i^2) - (\sum x_i)^2}}$$

The regression equation that we obtained in Example 1 is:
$\hat{y} = 151.35 + 1.399x$. Using this equation we calculate $\hat{y}$ for x=42 and obtain the value of 210.1.

The critical value for t is found in the t-table. For this example, $t_{\frac{\alpha}{2}}$ with (n-2) degrees of freedom = 2.179.

The standard error, s_e, which we obtained in Example 2 is 19.48.

To calculate $\bar{x}$, $\sum x^2$, and $(\sum x)^2$, press VARS, select **5:Statistics**. Highlight **2:** $\bar{x}$ and press ENTER ENTER. Notice that $\bar{x} = 42.07$. Press VARS again, select **5:Statistics**, highlight $\sum$, select **1:** $\sum x$ and press ENTER ENTER. So, $\sum x = 589$. Press VARS again, select **5:Statistics**, highlight $\sum$, select **2:** $\sum x^2$, and press ENTER ENTER. Notice that $\sum x^2 = 27253$.

Now, calculate the margin of error, E, when $x^* = 42$ using the formula for E shown above.

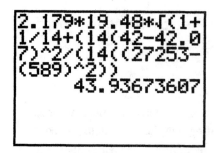

The prediction interval for the y-value at x = 42 is $\hat{y} \pm 43.94$. Since $\hat{y} = 210.1$, we calculate 210.1 ± 43.94 to obtain the lower bound of 166.16 and the upper bound of 254.04. These are the lower and upper bounds of the 95% prediction interval for the cholesterol level of a randomly selected 42-year-old female.

Section 12.3

▶ Example 1 (pg. 731)   Testing the Requirements of a
One-Way ANOVA

The four requirements for the One-way ANOVA test are listed in your textbook on pg. 730. The first two requirements state that the data must be obtained using simple random sampling techniques and that all samples must be independent. The third requirement states that the populations must be normally distributed. This requirement can be validated through normal probability plots of each of the samples. The final requirement, that population variances are equal, can be tested using the sample standard deviations. The criteria that we will use is the following: the largest standard deviation must be no more than two times larger than the smallest standard deviation.

Enter the sample data into **L1**, **L2** and **L3**. We will check for normality using a normal probability plot. To set up the normal probability plot, press **2ⁿᵈ [STAT PLOT]** . Press **ENTER** to select **Plot 1**. Highlight **On** and press **ENTER**. Set **Type** to the normal probability plot which is the third selection in the second row. Press **ENTER**. Set **Data List** to **L1** and **Data Axis** to **X**. Next, there are three different types of **Marks** that you can select for the graph. The first choice, a small square, is the best one to use.

Press **ZOOM** and select **9:ZoomStat** and **ENTER**.

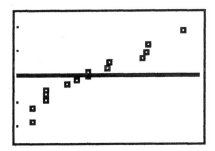

This plot is *fairly* linear, indicating that the data generally follows a normal distribution.

Repeat this process for the sample data in **L2** and **L3**. All three plots are fairly normal so the requirement of normality has been satisfied.

The next requirement involves the sample standard deviations. To obtain the standard deviation of the data in **L1**, press **STAT**, select **CALC** and **1-Var Stats** and type in 2nd [L1]. Press ENTER. The standard deviation for the data in L1 will appear on the screen (sx = 9.637). Repeat this process to obtain the standard deviation of the data in L2 (14.380) and L3 (9.414). Calculate the ratio of the largest standard deviation to the smallest standard deviation: 14.380/9.414 = 1.528. Since this ratio is less than 2, the requirement of equal population variances is satisfied.

> ▶ Example 2 (pg. 732) Using Technology to Perform
>   One-Way ANOVA Tests

This example is a continuation of Example1.

To test the doctor's claim that the mean HDL cholesterol levels for males in the age groups 20-29 years old, 40-49 years old and 60-69 years old are different, the test of hypothesis is: $H_o : \mu_1 = \mu_2 = \mu_3$ vs. H_a :at least one mean is different from the others.

In Example 1, we showed that the requirements of the One-Way ANOVA test have been satisfied. We also entered the data into **L1, L2,** and **L3**.

To run the hypothesis test, press **STAT**, highlight **TESTS** and select **F:ANOVA(** and type in 2nd [L1] ▯ , 2nd [L2] , 2nd [L3].

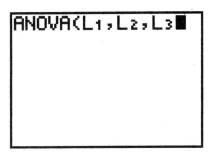

Press **ENTER** and the results will be displayed on the screen.

```
One-way ANOVA
 F=.592976295
 p=.5584614366
 Factor
  df=2
  SS=153.5
↓ MS=76.75
■
```

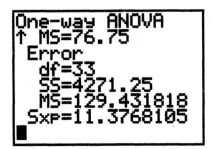

The output displays the test statistic, F = .593, and the P-value, p = .5585. Since the P-value is greater than α, the correct decision is to **Fail to Reject** the null hypothesis. The data does not support the doctor's claim that there is a difference in HDL cholesterol level among the three groups of men.

The output also displays several other pieces of information. This information can be used to set up an Analysis of Variance Table (similar to the tables given as output in Minitab and Excel).